H. Collins served as a Royal Marine in 40 Commando. He later worked again in Iraq as a security contractor, where he spent five years in the 'triangle of death', and Afghanistan; he was in Fallujah in 2005, and Ramadi until 2010. From 2011 to 2014, he worked providing maritime security in the IRTC (International Recommended Transit Corridor) through the Gulf of Aden. In 2014, he found himself back in Kabul, working as a team leader with civilian security contractors responsible for the diplomatic security of the Japanese Embassy. He was there in 2021 when one of the largest ever airlift evacuations took place, from Kabul.

www.lto.org.uk

CW00549190

This book is dedicated to:

Daryl Fern, a leader, mentor and warrior in every single sense of the word. You kept me alive and taught me some of the most valuable lessons in this life. I would give anything to take one more drive into the badlands of Ramadi with you. I miss you so very much.

Luke Griffin, the happiest, most upbeat guy in Afghanistan. You went out fighting. There is no higher honour than that of a man who moves towards the sound of gunfire, and Luke had the warrior spirit in abundance. We all miss you and that smile.

Nicky Collins, you never asked me to quit, and you never stopped believing in me. For me that will always be enough. I love you and thank you for your unwavering support throughout the years and the war zones. A life with you and our children is all I want.

Alexa and Jaxson Collins, dream big! And never let anyone tell you that you can't!

LAST
TEAM
OUT
OF
KABUL

Surrounded by the Taliban

H. COLLINS, WITH GRAHAM SPENCE

First published in 2022 by Ad Lib Publishers Ltd
15 Church Road
London SW13 9HE
www.adlibpublishers.com

Text © 2022 H. Collins

Paperback ISBN 978-1-802471-11-3
eBook ISBN 978-1-802471-13-7

A CIP catalogue record for this book is available from the
British Library.

Every reasonable effort has been made to trace
copyright-holders of material reproduced in this book,
but if any have been inadvertently overlooked the
publishers would be glad to hear from them.

Printed in the UK
10 9 8 7 6 5 4 3 2 1

Glossary

c. – colloquial/slang term used by Royal Marines

AQI	al Qaeda in Iraq
AK-47	*Avtomat Kalashinikova* assault rifle, also known as a Kalashnikov or AK
AR-15	a lightweight, semi-automatic ArmaLite rifle
B6	a grade of ballistic protection for vehicles which protects against high-power rifles – B7 provides the highest level of protection, from armour-piercing bullets
Bergen	a 100-litre rucksack, standard issue for Royal Marine Commandos
bootneck (*c.*)	a Royal Marine
BPT	Brigade Patrol Troop, a crack surveillance and reconnaissance troop of the Royal Marines
chalk	a group of soldiers who deploy from a single aircraft
CLR	Commando Logistic Regiment
CP	close protection
CTCRM	Commando Training Centre Royal Marines
CWS	common weapon sights
DFAC	dining facility (US military)
EFP	explosively formed penetrator, to penetrate armour, including that of tanks
EOD	explosive ordnance disposal
EOTech sight	an advanced gun-sight developed by US company EOTech
ETA	estimated time of arrival
FFD	first field dressing
foxtrot	military radio jargon for 'moving', on foot
Gucci kit (*c.*)	Royal Marine term meaning top-notch equipment
headsheds (*c.*)	officers in charge
helo (*c.*)	helicopter
HLS	helicopter landing site
HMG	heavy machine gun
IBG	individual bodyguard
icers (*c.*)	Royal Marine term to describe being freezing cold

GLOSSARY

IDF (*c.*)	indirect fire, such as mortars or rockets
IED	improvised explosive device
IRTC	Internationally Recommended Transit Corridor
Javelin	a portable anti-tank missile system
JDAMs	Joint Direct Attack Munitions, guidance kits that convert unguided bombs into precision-guided bombs by means of an inertial guidance system and a global-positioning system (GPS)
L96A1	the British Army's standard sniper rifle
LAW	a light anti-tank weapon, used by the infantry
long (*c.*)	a Royal Marine term for an assault rifle
M4	assault rifle widely used by US troops
Minimi machine gun	a comparatively lightweight, belt-fed machine gun designed to be used by dismounted infantry
MREs	Meals Ready to Eat, food packs issued to US soldiers
muezzin	a public crier who proclaims the regular hours of ritual prayer from the minaret or roof of a mosque
NAAFI	Navy, Army and Air Force Institute, a company created by the British government to run recreational establishments for the British Armed Forces, and to sell goods to servicemen and their families
NBC respirators	nuclear, biological and chemical gas masks
NBC	nuclear, biological and chemical, as in NBC preventative procedures, for example
NCO	a non-commissioned officer
NVDs	night-vision devices
NVGs	night-vision goggles
op (*c.*)	a military operation
Pashto	a national language of Afghanistan (it is also spoken in Pakistan)
PHC	primary health clinic
PKM	a 7.62-mm belt-fed machine gun
PRMC	Potential Royal Marines Course
QRF	Quick Reaction Force (US military)
RPGs	rocket-propelled grenades
RSM	Regimental Sergeant Major
run ashore (*c.*)	Royal Marine slang for a 'night out on the town'
SA80	standard-issue assault rifle for British Armed Forces
SAS	Special Air Service
SBS	Special Boat Service
SET	security escort team
VBIED	vehicle-borne improvised explosive device
wet (*c.*)	Royal Marine slang for a cup of tea
WMIK	weapons mount installation kit – a modified Land Rover with a reinforced chassis and mountings on roll bars to run heavy-calibre weapons
yomp/ yomping	Royal Marine slang for a long-distance march in full kit/doing such a march

Contents

Prologue

'You are wearing the watches. But we have the time'

Taliban commander to an American general

In my line of work, part of preparing for grave situations is writing what we call a 'death letter'. It's where you say your final piece – your last words – to your loved ones, in the event of your death.

The last time I had written one was in Ramadi, Iraq, where getting hit by fanatical insurgents was a daily occurrence.

After harrowing nights on the rooftop of our security headquarters in Kabul, watching and waiting as the Taliban took over Afghanistan, I had written another farewell note to my wife Nicky. At first it was a quickly punched-out text – a rather terse, 'I love you, tell the kids I love them, move on,' sort of thing.

After a week of not knowing if I would make it, the death letter morphed into a fairly heavily worded document that exceeded the 700 characters allowed in a text. It then became an email so I could fit everything in, and I saved it as a draft. If the Taliban attacked us, I could simply open my phone, hit 'send' and it would get to Nicky, hopefully before I was gone.

Re-reading the 'death' email in the small room assigned to me as team leader of the Japanese Embassy's security unit, I watched the bar flash on my mobile phone, inviting me to write more. I was already upset, but if these were to be my last words, they had to be the most meaningful I had ever written. My children would read them in years to come and I had to let them know the answers to anything they might have wondered, including what had gone on in Afghanistan, what I was doing there and how I had lived my life. I also gave some life-experience advice and, most importantly, an apology for being away from home on contract so often.

My hands shook a little as I held the phone. What I'd written was enough. It was all in there and I did not hold anything back – good or bad, right or wrong. It was the truth poured out from deep inside me. I closed the phone and checked my bag once more: pretty much a pointless task, but it stopped the emotion of what I had written from overcoming me.

Soon the Taliban were – apparently – going to escort us to Kabul Airport, where hopefully we would fly home. It was to be the final extraction of the Japanese Embassy. But the bombshell news was that we were all going to be disarmed, and we would not be using our own vehicles. The Taliban would drive us. Without weapons, we would be totally at their mercy.

One of my team members, Bry – an eighteen-stone giant of a man – summed up our reaction in two words: 'Fucking hell!'

The Taliban arrived that morning and popped open the gates, and the Gurkhas employed by the embassy as static guards were the first to make their way towards the armed and menacing bearded men about to search them. They began a systematic examination of each and every bag, rummaging deep and pulling out almost all of the contents. The way I saw it, the search was to confirm one thing – that we were not armed. The queue grew smaller and smaller, until our security team members were the only ones still to be searched.

We shuffled slowly forward, past abandoned bags and personal belongings scattered on the ground. It felt surreal moving unarmed towards the Taliban and it went against every natural instinct I had as a former Royal Marine Commando. In 2002 I had been part of the 'spearhead', the invasion force that had routed the Taliban from Afghanistan soon after Osama bin Laden's al Qaeda had flown planes into New York's Twin Towers. Now I was approaching the guys we had been fighting against for twenty years with nothing more than a backpack and my bare hands.

One of them was speaking in English. He seemed angry but I couldn't make out what he was saying. 'Great!' I thought. 'I get the angry one.'

As I got closer, I could hear his rant. 'If you are an engineer, if you are a scientist, if you are a doctor, you can be in Afghanistan. But if you are a soldier, you cannot be in Afghanistan.'

That's us screwed, then.

In front of me was one of my team. He moved nervously forward for the search. His backpack was open and the Taliban soldier bent down, pulling out bits and pieces, and unclipping the ballistics helmet attached to the outside of the pack, throwing it to the side.

He then pointed to the body armour that my guy wore.

'Take off!' he commanded.

My team member started pulling at the Velcro and removing the armour. It was obvious the Taliban wanted us to leave with nothing other than what we stood up in. No kit was going to make it out and certainly no tactical kit was going to pass through that gate.

Witnessing this, I realised he was going to tell me to do the same, so I began removing my body armour. The smell of stale sweat wafted past my nose as I lifted the ballistic vest over my head, stripping down to nothing more than jeans and a grey T-shirt.

The searcher turned his attention to me, patting me down in a body search. I was now officially closer than I had ever been to the Taliban, literally face to face. I could smell the cigarette smoke on him. As he stood, we locked eyes. His lips parted in a thin grin. This was his moment in the sun and he knew it.

Satisfied I was unarmed, he pointed to where everyone else was standing, including my team. I had made sure I was last to go through the body search, in case anything went loud at the last minute, and I still had the chance of getting to our weapons and fighting towards a getaway car. But now I was walking over to the enemy with no options left whatsoever. My team watched me and I could see the worry etched on their faces.

We were placed in a large concrete open area, boxed in with high walls on three sides. One way in, same way out. A pure killing zone, if ever I had seen one.

For a few moments we stood in silence. The Taliban had complete and utter control, and we all knew it. A feeling of dread washed over us. The joking and dark humour so prevalent among my team – most of whom were battle-hardened veterans – was gone, replaced only by an empty acceptance.

Then I heard the sound of vehicles, accompanied by a fair amount of shouting. Something was about to happen that had got the Taliban body-searchers even more animated.

I was unsure what had sparked this sudden flurry of excitement. Then three pick-up trucks, loaded with Taliban, roared into view. On the back of each truck was a PKM, a 7.62mm belt-fed machine gun, all pointing at us.

This was it, the moment they could kill us without any resistance whatsoever.

My heartbeat started going through the roof. Then two Taliban fighters jumped down and began filming us on their phones. I had seen this many times before. It was an execution video – or a 'snuff video', as the media called them.

The men talked as they filmed and slowly made their way towards us, ensuring that they got everyone's faces. I pulled my baseball cap further down. I didn't want my kids to see my face after I had been killed. It's traumatic beyond description seeing someone you care about dying violently, as the victim's face can have a really strange expression in death. I knew that first-hand, having seen it often. It's hard to get that horrific image out of your head, and having loved ones see that would be truly disturbing. Above all, I wanted to save my family from such anguish.

One of the video men passed by in front of our team, making sure he soaked up the moment of victory and captured it on his phone. His commentary in Pashto appeared to be unemotional, as if he was reading off facts or statistics. Perhaps he was counting to get the exact death toll. It would be high.

I stealthily pulled my phone from the back pocket of my jeans and unlocked it, opening the draft email for Nicky. I hovered my thumb above the send button.

At that moment I felt two overriding emotions: fear and love. Never before had such simultaneous sensations surged so powerfully within me. The fear was for my life, but the love for my family was overwhelmingly stronger. It took me to a heightened place – a place of absolute clarity, where only truly important things could enter my thoughts. I realised at that instant that nothing else mattered – only love, and we are all connected by it. I was connected by love to Nicky and my children. I was not with them physically, but even so, I had the feeling that I was not alone.

Standing there, unarmed in the shadow of PKM machine guns manned by the enemy, and thinking of my family was a defining moment, internally connecting me to everything that was meaningful in life. It was an acceptance of who I was and what I had done on this Earth. I was at peace with myself, unshackled from fear. The Taliban surrounded us in large numbers, parading around and continuing to film on their phones. I knew exactly what would happen once the cameras stopped. I watched the machine gunner on the back of the closest pick-up. The second his finger went near that trigger, I would hit 'send' and that would be it. My final piece would be read by Nicky and my children.

Time stood still. The heat seemed to intensify and I could smell freshly oiled weapons each time one of the Taliban passed me. This was a sure sign that they intended to kill us, as the last thing they wanted was a weapon to jam, especially with this many targets.

My thumb never left that spot above the send button, as my eyes watched for any sudden movement, a clue, a combat indicator ...

For twenty years I had been at the sharp end of every major global conflict, from Kosovo to Iraq – some by chance, some by choice.

This is the story of how my team and I overcame our biggest psychological battle of all: being surrounded by a city full of Taliban fighters with no way out.

1

The Shire

Lincolnshire is two things: flat and boring. Growing up there is a painful process that I would not like to repeat, and thankfully won't have to.

I was a restless teenager, craving a life of action and adventure – both of which are distinctly lacking in the shire. In fact, as far as I was concerned, the only excitement was the roller disco held on Saturday nights at the local sports centre's skating rink. For me and my mates, this was the single shining highlight of the week. We were all avid skaters, and the disco was the perfect place to show off our skills and maybe earn a kiss or two off the girls watching from the sidelines.

Consequently, with not much else to do, my life revolved around skating. It was what I lived for, and I seldom went anywhere without wheels on my feet. I could cut across to most of the surrounding villages, and this was not only my obsession, but my main form of transport. I had no real interest in other sports and during rugby or football games in physical education classes, I would be the one standing shivering on the sidelines.

That all changed one Tuesday evening, when my parents took me to a careers exhibition. I would soon be leaving school and, as I had shown no academic inclination whatsoever, they were perhaps trying to give me some direction in life.

The exhibition was in a hall where stands promoted various job opportunities, while people waving recruitment brochures greeted us at the door. I was already bored, when suddenly I noticed a guy standing at a stall off to one side. He was impressive in every sense: physically towering over everyone, while his arms and chest bulged under a combat shirt that looked as if it was shrink-wrapped.

'Who the hell is he?' I wondered. And why did he have such an incredible aura – a sort of animal magnetism – surrounding him?

Gingerly, I headed over, wanting to see what he was doing among all the other brochure-wavers, but nervous about asking him.

He caught my eye. I must have looked like a vulnerable creature in a spotlight as he extended his hand.

'Hi, I'm Corporal Jenkins. Have you thought about joining the Royal Marine Commandos?'

I hadn't. In fact, I had barely even heard of the Royal Marines. Who actually were they?

He started telling me and I was instantly captivated, totally and utterly spellbound. He recounted some of his own experiences, ranging from far-flung tropical paradises to the icy highlands of Norway. He was everything I wanted to be: not only living the dream, but also getting paid to do so. That was what I wanted – that was the action and derring-do that I craved.

I will always remember that day. It's still illuminated in my mind in brilliant Technicolor. My life's journey began at that moment. No longer a bored, disgruntled kid hating life in the shires, I was filled with a new sense of purpose. That twenty-minute chat with Corporal Jenkins had irrevocably changed my destiny. A destiny that would catapult me into a life of danger and adventure around the globe: a life of fun, passion and excitement – but also of terror, horror and shock. The life of a frontline Marine Commando and security contractor in the hotspots of the world.

The effect was profound. Overnight, my frivolous obsession with roller-skating was channelled in a completely different and more serious direction: I was going to become a Royal Marine.

My biggest problem was my body. My mind was convinced I was going to become a warrior, but my body had not caught on to that yet and a lot of work was needed. Strength and fitness are the two most basic requirements for the three-day PRMC (Potential Royal Marines Course) I first had to pass, and I had neither.

I began my physical preparation with a simple training regime of running. Once I had finished my paper round after school, I laced up my trainers and went for a mile-long run. That was the idea, anyway, but reality soon hit hard. I could barely breathe after the first 500 yards, as my lungs were on fire. My leg muscles, which had never done anything other than walking and skating, seized in agony.

Fitness did not come easily. It was hard work. For a start, I was mildly asthmatic, and my small-framed body took a brutal hammering from what my mind was instructing it to do. But I never gave in to those muscles screaming at me to stop. Even at that age, I grasped that the body would persevere, as long as the mind was strong.

So I chipped away at my fragility, bit by bit, never surrendering. And slowly it started to pay off. After a couple of months, that torturous mile I struggled to run became a mile and a half, then one and three-quarters, and then two. I also began timing the sessions to see if I was within the Royal Marine cut-off time of 12.5 minutes for a 1.5-mile run. Usually, I wasn't.

I also started doing multiple press-ups, sit-up and pull-ups, as Marine fitness involves both strength and stamina. Selection criteria push candidates to the absolute limit of physical endurance, with gruelling assault courses, fast-running fireman carries and long-distance, full-kit marches known as 'yomps'.

As I got fitter and tougher, the dark, cold and wet Lincolnshire morning runs became bearable, as I began to consider them a challenge rather than a chore. I took it personally when rain and wind battered my bedroom window, as if nature was deliberately taunting me, trying to get me to admit defeat and stay in a warm bed. Instead, I got up and ran.

I started pushing myself harder and harder. Whenever I achieved a goal, my next step would be to ratchet it up a notch. I started running in heavy boots, or with a weighted backpack, conditioning my shoulders to carry loads, and my lungs and legs to stay the course. I was actively seeking discomfort – something that would stand me in good stead in later life.

Then, a strange thing happened. The harder I trained, the more I started enjoying it. My favourite exercises were pull-ups, press-ups and the beep test that measures VO_2 max, considered the most accurate yardstick of fitness. This test – continuous running between two shuttles – always made me nervous, but the sense of achievement after completing one was fantastic. To pass PRMC requirements, I needed to get to Level 11, which is pretty tough for a kid who not long ago thought going to a roller disco and kissing a girl was the pinnacle of achievement.

Unsurprisingly, schoolwork took a distant back seat in my life. I had always hated it, but by then I regarded classroom lessons as little more than rest-and-recovery periods between my morning and evening training sessions. I had no interest in the academic world – I was going to be a Royal Marine and no one, including my long-suffering teachers, was going to tell me any different. That was my only goal in life.

Once I was ready physically, I arrived at the military careers office in central Lincoln, nervous but determined. I was then directed to a naval officer who was to be a mentor during my application. As I was only sixteen years and nine months old – the youngest legal age for a recruit – my parents had to give written permission.

My three-day PRMC application was approved, leaving me a handful of weeks to finish off my preparations at home before catching a train from Lincoln to the Lympstone Commando in East Devon.

Before me lay a series of challenges to see if I had the mental and physical ability to be considered for the next stage: passing a brutal, thirty-week selection course to earn the right to wear the coveted green beret. It was, in essence, a preliminary job interview, but one completed with sweat rather than pen and paper.

My hard work paid off and I passed the PRMC. I was then invited to do the full Commando course, widely regarded as the toughest selection programme outside the special forces units.

I would start on 16 December 1997. The winter intake.

2

Lympstone Commando

Nestling beside the estuary of the River Exe is the Commando Training Centre Royal Marines (CTCRM), the first base for any recruit wishing to join this elite rapid-reaction regiment. As the train slowed down for the village of Lympstone, the knot that had been in my stomach all day tightened.

I was there for real. Providing I made it through selection, this would be home for the next eight months.

'Move, move, line up here!' shouted a man with a stick under his arm. He was impeccably dressed – shirt crisp and pressed – and standing perfectly still and straight as he barked orders at a hundred miles an hour. I quickly moved as instructed.

Corporal Weston was the chief instructor for my group, 735 Troop. He was the perfect Marine, both physically and mentally – not too big, not too small, intelligent and quick-witted. He had an answer for everything and, as a former combat Marine, had seen and done it all before. At the Commando Training Centre, he was God.

On our first day, we were issued military equipment and shown how to look after it, as well as instructed on how our lockers must be kept; the importance of personal hygiene and spotlessly clean living quarters was drummed into us so often that it became second nature.

Marine life at CTCRM basically revolved around three locations: the assault course, the indoor gym and the fabled wildlands of Dartmoor. The assault course and gym built up peak fitness and strength, without which it would be impossible to survive training on Dartmoor. The rugged moorland is a deliberately chosen testing ground: a miserable, boggy, bitterly cold and unforgiving – yet strangely beautiful – hellhole. Carrying full kit on one's back is little short of crippling in the starkly harsh landscape, as the hills stretch forever, getting steeper all the time. It's a supremely challenging environment: continuous hiking with heavy packs in the day, as well as having to plan and execute tactical situations at night, while utterly exhausted, ruthlessly eliminates those not cut out to be Marines.

That was the whole point. If you couldn't handle being freezing, disgruntled, hungry, miserable and sleep-deprived, this was not for you. I didn't know it then, but those severe experiences would prove crucial to my survival in future escapades.

Doing Commando training at such a tender age was probably a good choice for me. My young body healed well, and because I didn't know any better, the shouting and abuse continuously hurled at us didn't bother me. If anything, I regarded it as normal, whereas some of the older guys who had more experience of civilian life hated the bullshit. They were men and wanted to be treated as such, whereas I was still very much a boy and could handle it.

Thirty weeks is a long time – in fact, the lengthiest basic training programme of any NATO combat infantry unit. And it seems even longer when you're wet, dejected and tested to your limits around the clock. Each week was harder than the previous one and our kit got heavier. My shoulders started to bulk up, partially due to growing into manhood, but also from the ridiculous amount of weight I carried in a Bergen, week in and week out. A Bergen is a hundred-litre rucksack which can pack a lot of kit. Usually, it takes two men to lift one onto your back, and once it's on, you'd best not fall over. You won't get up. The straps cut deep and your shoulders burn fiercely from the load. Then when you start moving, your buckling legs also start taking strain, as three plodding steps are equivalent to one normal pace – and when you hit the endless hills of Dartmoor, everything is on fire: shoulders, back, arms, legs and lungs.

Marines call hiking with a Bergen 'yomping', an iconic word in the arcane bootneck language. We're called bootnecks because, initially, Marines were a Royal Navy captain's personal bodyguards and slept in front of his cabin. During a mutiny, the crew would first have to slit a sleeping Marine's throat to get to the skipper. Consequently, Marines wrapped the tongue of a leather boot around their necks so they could wake and fight back. Other typical Marine words are 'wet' for a cup of tea; 'icers' if you're freezing cold; and a night out on the town is a 'run ashore'.

Yomping is massive in the Marines. It's a key part of the gruelling endurance tests, and it breaks people – physically, spiritually and mentally. When on long yomps and exhausted beyond description, I would always repeat the insults I got from some schoolmates who knew I was trying out for selection.

'You're not fit enough!'

'You're not strong enough!'

'You won't get in the Marines!'

It became a mantra. The repetition of those lines fired me up; I was desperate to prove them wrong. Every step was literally a step of defiance to my body that wanted so desperately to stop.

Also, I would let my mind roam freely during a yomp, thinking about anything except the pain I was enduring. I would recall the local girls I had met and wonder what they were doing, or who they were with. Or what my schoolmates were up to, and the parties they would be going to while I was on Dartmoor, struggling with a huge rucksack in a wintry gale.

Joining up in winter is particularly savage, as you simply can't get warm on those bleak windswept hills. No matter what you do, you freeze to your core. And don't expect any sympathy from the instruction team. On one long punishing yomp, I briefly leant forward for a couple of seconds, easing the tension on my frozen shoulders and resting the Bergen on my back. That blissful respite lasted all of two seconds, before Corporal Weston started berating me.

'Keep moving! You think this is cold?' he shouted. 'This is bugger all, lads. Wait until you get to Norway.'

He could ruin your day with a single glance, but most of all he could destroy morale in one short, sharp sentence. No matter how bad things were, he would tell us they would get worse. Breaking our spirit was his hobby. Or so it seemed at the time. Nowadays, we are good friends.

Apart from getting us Olympic-level fit, the first fifteen weeks of recruit training also taught us the basics of military life, after which we moved onto sharpening fighting skills, amphibious warfare, climbing and rope work – until the final stage: the dreaded Commando course.

It was in the first phase of training that I developed a nagging ache in my left foot, which bugged me but I thought was nothing serious. I found that my young body recovered quickly from any serious physical abuse. This would be no different.

I was wrong. During the second phase, just four weeks away from starting the Commando course, disaster struck. I swung out of bed one morning and an agonising pain knifed through my foot. It could take no pressure whatsoever. Even touching it lightly hurt.

I managed to hobble about twelve yards before pulling up. I looked around, almost panicking. That morning we were scheduled to go on a speed march to the shooting range for rifle practice. How would I be able to do that?

As usual, I got no sympathy from the training team. 'Collins, get a fucking grip of your body!'

I didn't know what the pain was, but I knew it was serious. Walking, or even standing, was out of the question. I was helped into the back of a support vehicle – the 'jack wagon', as we called it – that collects recruits unable to keep up.

The training team organised a lift back to CTCRM, moving me from the misery of Dartmoor to the medical centre at camp. That was the last place I wanted to be. It's where aspirations of becoming a Marine are shattered into a thousand pieces – the graveyard of any recruit's dreams.

X-ray scans revealed my worst nightmare. My foot was broken in two places. I was devastated. Totally destroyed.

The medical guys outlined my rehab programme, but I barely heard them. I was inconsolable. My world, and everything in it, was gone. It felt like it had all been a massive waste of time and effort.

I was sent to Hunter Troop, where injured Marines are restored to full health and then re-inserted into another Troop at the stage where they had been forced to leave. In my case it was week twenty-two, just eight weeks away from earning the coveted green beret. So near, and yet so far.

I was convinced that any chance of becoming a Marine had been well and truly dashed. This was mainly because of the many stories told about Hunter Troop, such as it being only a place for skivers and those unable to cut it. Once admitted, you never became a Marine.

Fortunately, most of those stories are total bullshit. It didn't take me long to discover that Hunter Troop did have people genuinely injured and absolutely determined to get back. I met some truly fantastic lads there.

I began my rehabilitation with a physical training instructor called Sergeant Sylvester. From the word go, he was amazing. He knew how I felt and took his time with my recovery, making sure my head was still in the game, even if my body wasn't. I wanted my foot healed yesterday, but he was a consummate professional. My foot had to heal properly, and he would not be rushed into sending me back to my Troop. It was superb how the rehab team wanted me to get back out there and succeed.

My mornings were spent swimming so I wouldn't have to put pressure on any broken bones, while still using my muscles and maintaining cardiovascular fitness. In the afternoons I worked out with a resistance band for strength and after that my foot was iced in a freezing inflatable sock. I would sit there on the floor mats in the middle of the gym, jealously watching other recruits, who had recovered fully, pounding away on the treadmill. I would never take the gift of a fully functioning body for granted again.

This daily routine was my new normal and I started making progress. After a while, I could finally stand on the foot. Not needing crutches was amazing, even if it was for just a few steps.

As well as physical healing, Hunter Troop also worked on preparing me mentally for the final Commando phase. I needed that, as this section is where most people fail. Everything recruits do, everything they learn at Lympstone, climaxes in this absolutely crucial part of selection. The tests are a brutal physical and mental evaluation, and they come one after another in rapid succession. I have seen super-strong and super-fit guys fall to pieces in those final weeks. There is no time to recover, no time to gather one's senses.

The key objective is to drive potential Marines well beyond any normal limit of human endurance. I had, to a lesser extent, touched on this while training for the PRMC, when at first I couldn't even run 500 yards. But I learnt, maybe without fully understanding it then, that we all have a mother lode of innate mental and physical steel. However, in order to find that secret source, we have to tap the depths of our inner selves. It's something few people wish to do, as it demands peaks of pain and troughs of exhaustion infinitely more intense than mere fatigue. The human body and mind can perform at far higher levels than most people believe. Once someone grasps that, they can do anything, if they are willing to pay the price.

After several months, I was discharged from Hunter Troop and sent back to the selection course.

I was about to find out just what price I was willing to pay.

3

The Commando Test

After ten days of what's called a 'final exercise' of living rough in the wilds of Dartmoor, enduring never-ending yomps and routine sleep deprivation, we returned to camp to face the four Commando tests.

Consequently, my legs were already sore when I woke on the morning of the first test: the Nine-Mile Speed March. Speed marching never really bothered me. Sure, nine miles was tough but for me it was a case of rhythm – just staying with the pace. The other three tests were far worse, in my opinion, and I was already nervous about them.

It was March, and still cold, as we set off towards the start line just beyond the camp gates. Webbing straps were already rubbing into my shoulders, but by then I had mostly learnt to ignore the pain. It was just another Marine rite of passage. Once the Troop was present, the clock started. We had ninety minutes to do the loop back to the camp in full fighting order, averaging a mile every ten minutes.

After a while, a few guys slowed. This was a problem if you were marching behind them, as it slowed you down, too. Waiting behind lagging guys means you won't be able to catch up with the rest of the Troop.

When we reached the last mile post, I knew I was going to make it. We arrived at CTCRM and were drummed into the camp down the main road. Everyone has to clap you in, so there are literally hundreds of recruits lining the road. It makes the hairs stand up on the back of your neck when marching past like that, and this officially marks you as the 'Kings Squad', the most senior troop on camp.

But this was just the first of four tests. And to get a green beret, I had to have a clean sweep.

My feet were sore and I tended to them that evening, cleaning and re-dressing a few blisters in preparation for the Endurance Course. This was not my strongest event. I had failed it in the rehearsal by two seconds, and it had crushed me. It involved a fast two-mile run with twenty-two pounds of kit and rifle across Woodbury Common, wading through waist-deep pools and tunnels half-filled with water, and finally swimming through a narrow underwater tunnel.

But the so-called 'two-mile' course itself was nothing of the sort. We first had to run four miles just to get to the starting line on Woodbury Common, and once we had finished, it was another four-mile semi-sprint back. So, in reality, it was ten miles in full, thoroughly waterlogged, combat kit and mud-caked boots.

I wasn't alone in my apprehension, as a lad called Ward had shat himself – literally – every time on the run back. It became the Troop joke and he loved it. For some reason, he seemed to enjoy shitting his pants. Weird as this sounds, I think we all got to the point during the

Commando phase where nothing else mattered but finishing, and if you didn't have time to stop for a crap, what did you do? You did it in your pants, or in Ward's case, your combats.

We set off at one-minute intervals. I always aimed to catch someone ahead, knowing it earned me a bit of a time window. But what made this test even more of a bitch was that our kit and rifles snagged on every obstacle imaginable in those intensely claustrophobic water tunnels. And we had to make sure our wet rifles still worked, because at the end there was a marksmanship test. With heart and lungs pumping like a F1 engine, we had to steady ourselves and fire ten accurate shots at a twenty-five-metre target simulating 200 metres.

I passed. I don't recall my time, but I was within the limit and my bullets hit the target.

Only two more tests to go.

Up next was the Tarzan assault course, which was short, sharp and horrible. It's a mixture of formidable obstacles, climbing nets, zip wires and no room for error – as we discovered while watching another troop doing it. One of their lads got it wrong and fell from some height, bouncing off the floor with a real thud. The most sympathetic advice from his training instructors was, 'Get up, you useless bag of shit.'

This was my first experience of the Tarzan, and as I have a fear of heights, watching someone plummet to the ground just before I was going to attempt the mother of all assault courses didn't exactly put me in the right frame of mind.

My name was called and I climbed up into a large tower from which I had to zip-wire down to the first part of the high obstacle course. The tower dominates the skyline at CTC and, trying not to look down, I threw myself off it, zipping down the wire at an exhilarating speed. I hit the landing platform, and then had to navigate my way through a maze of climbing nets and high-strung ropes – hence, the name 'Tarzan'. Being small and agile was an advantage, and pulling up my body weight of ten stone was never a problem. Despite my fear of heights, I finished the Tarzan in a respectable time.

I then sprinted down onto the bottom field for the assault course. The first obstacle was a stretch of water that I had to jump across while carrying twenty-two pounds of webbing and a weapon. It had to be a two-footed take-off and a two-footed landing, and I reckoned that if I did it at full speed, I might just make it.

That's when it happened. As I took off, a white-hot shaft of pain sliced through my left foot. The same pain I had experienced before. The same foot.

'No!' I almost screamed. 'Not now! Please God, not now!'

I was about seven obstacles from the finish line. I ran painfully to the next section, supremely grateful that at least the foot could bear my weight. Fortunately, this challenge was a rope suspended between two frames and didn't involve using my feet. I dragged my body across.

'Almost there,' I told myself.

The uphill section of the assault course is a real ball-breaker, and can only be done pretty much at walking pace, which also helped me. Next hurdle was climbing over a wall. Using my right foot to thrust myself upwards, I managed to grip the top bricks, ripping my fingers in the process. Then it was a half-limp, half-jog to the tunnels. Unlike the Woodbury Common course, these were not awash with water. Instead, they were filled with broken bricks – great fun for kneecaps and hands. If you were not bleeding before, you definitely would be by now.

Crawling through tunnels is usually okay for me, but this time I battled painfully through them. I emerged into daylight, gasping for breath and feeling a little sorry for myself. It was looking as though I wouldn't make it.

The final challenge was climbing a thirty-foot wall with ropes – something I am normally good at, but not with a crippled foot. Trying not to think of falling, I grabbed a rope and started hauling myself up. My foot throbbed incessantly, but the end was in sight. As I pulled myself over the top, I screamed: 'PO55016F Collins, Corporal.'

I had no idea why I had to shout my name and number, as I had been with my instructors for twenty-eight weeks. I'm pretty sure they knew who I was.

The instructor called out my time. It was a pass! I was relieved beyond words, but didn't show it. You never showed emotions with the training team. There were no high fives or anything like that. It was more of a case of a casual nod and a 'right that's done, what's next?' attitude.

We went back to our accommodation to shower and change. Three challenges down and one to go. All that lay between me and my goal of becoming a Royal Marine Commando was the Thirty-miler.

This is the toughest of the Commando tests. It's cemented in Marine folklore, with horror stories passed from generation to generation of recruits. I had heard them all: lads passing out, breaking legs or ankles a mile from the finish, getting lost, suffering from chronic hypothermia and some even dying. It was not meant to be easy. If it was, then anyone could be a Marine.

That evening, trucks took us to Okehampton Battle Camp at the edge of Dartmoor, where we would spend the night. It's a bleak, cold, windy and miserable place, and the dilapidated camp itself looks like a set for some spooky Second World War movie – an atmospheric starting line for the final showdown.

I was apprehensive about my foot. Although it hadn't got any worse, the pain was still there – a deep, dull ache. I asked if anyone had painkillers, and the Troop held a quick whip-round. The guys had remedies for everything: tape for blisters; Deep Heat for pulled muscles; Dioralyte for dehydration; Nurofen for pain. We always helped each other out – something we had been taught from the first day. Unselfishness was one of the many Commando values hammered into us.

Soon I had a fistful of pills. I threw some down my neck, and the effect was almost instant: I could place my full weight on the foot. This was the magic I needed, and I put another twenty-eight pills in a pocket, ready for the Thirty-miler.

Before turning in, we were given maps and safety kit, as Dartmoor is a cruel mistress if you get lost or separated. Our planning had to reflect that. Spirits were high: this time tomorrow, for most of the guys eight months of hell would be over. For me it would be more than a year, taking my injury into account.

However, I slept badly that night, dreaming crazy stuff that was probably the result of a combination of nerves, painkillers and high-energy drinks playing havoc with my system.

We were up before dawn and given some lukewarm food that quickly turned ice-cold in the ancient mess tins it was served in. The tea and coffee were just as hideous. I had some High5 energy sachets, which I used to swill down three more painkillers.

Our team of eight – with two extra, bulky safety Bergens that we shared carrying – walked to the starting line with hefty thirty-two-pound loads on our shoulders. Dark clouds gathered across the moors. The only other sign of life was a flock of sheep huddled together for warmth.

The knot in my stomach tightened. I was on the brink of becoming a Royal Marine Commando. Or not. Standing on a cattle grid, about to take off in the driving rain and a little buzzed by painkillers, I knew this was where I was meant to be. Excited voices cut through the darkness of that windswept morning and my confidence grew a bit, as the lads in the Troop rallied around me.

'Hey H, you got this far,' said one, pointing at my foot. 'This is the last thing.'

The others nodded. 'All that stands between us and a green lid is thirty miles.'

And, of course, eight hours – the time limit for this final Commando test.

'Good morning, gentlemen.' Our training team had arrived and the Troop boss began a rundown of what would happen over the next eight hours.

'Today marks what could potentially be your last day as a recruit,' he said. 'Upon completion of this fourth and final test, you will be awarded the green beret marking you as a Royal Marine Commando.'

I don't think any of us needed reminding of that as we stood in the freezing cold. For me, personally, my entire life had been focused on it since my last year of secondary school.

The law of physics states that what goes up must come down. But not in a Marine yomp. It's just one uphill after another. When you think you've reached the top, you're greeted by more uphills.

By the time we arrived at the first ten-mile checkpoint, my quad muscles were so full of lactic acid from all that pumping that they were as tight as drums. But my foot was holding up, although I was popping painkillers on the hour. At the checkpoint, I swallowed three more after gobbling some food.

Twenty miles to go. Then dense clouds closed in and one of the guys, who had previously worked as a farmhand on the moors and knew the weather patterns, shouted over the wind, 'Here it comes, boys.'

He was right. The mist turned into drizzle that soaked our kit, making it heavier and us even more miserable, if that was possible.

By then we had passed the threshold of mere pain. This was deep into mind games: how much more could one endure of such extreme conditions of discomfort, fatigue, misery and outright torture? How much longer could one keep going in such adversity?

Or, more succinctly, how badly did we want to become a Marine? To get that green beret? That was the answer our instructors were looking for.

To take my mind off the awfulness of the present, I tried to think of other things. I mentally clicked off, as I daydreamed of girls that I had liked but not bothered to pursue because of my single-minded obsession with becoming a Marine. I knew I had hurt one girl's feelings, in particular, when I left, and I regretted it. I was pretty harsh and cut her off, leaving her upset, and that weighed heavily on me. Maybe I could try to sort it out later. Maybe I could repair that damage.

That brief subconscious interlude made a mile or two pass more easily, but that was coming to an end. The pain in my foot had started to worsen. I popped another pill, hoping it would numb the ache enough to reach the next checkpoint.

The one brief glimmer of cheer was that, at the halfway mark, the foul weather had finally started to ease. We had a compulsory ten-minute break to eat, drink, and adjust boots and kit. A medic then gave everyone the once-over, and as he knew about my foot, he asked how it was holding up.

'All good. I'm smashing these in,' I said, proudly showing him an empty packet of Nurofen.

He was aghast. 'Fuck me, you're kidding!'

'It's okay,' I said, smiling like a smart-arse. 'I have another pack.'

'Hand them over now! Those things will kill you on something as extreme as this. You'll pickle your liver.'

I felt like a scolded child, as I gave him the pills.

'Okay everyone, gear up and let's get moving,' barked an instructor.

Miserably, I gathered my kit. In the mere eight minutes I'd had it off, it seemed to have doubled in weight. My shoulders were a mess, and I could feel the sticky warmth of blood where my Bergen straps had chafed through skin on the right-hand side of my back.

The CTCRM commanding officer joined us for the next fifteen miles, jog-walking easily along us, even though he was much older. Major 'Buster' Howes was an absolute legend and did a Thirty-miler in full kit every two weeks with recruits attempting the final Commando test. I liked him – in fact, all of us in training regarded him as a father figure. He later went on to become Commandant General of the Royal Marines.

I was struggling. It was my turn to carry the heavy and awkward safety Bergen, as well as my thirty-two-pound rucksack for the next few miles, which seemed to make my foot worse with every step. The pain was intense, doubly so as the pills had worn off and the medic had confiscated the rest of my supply.

I arrived at the twenty-mile checkpoint, my breath rattling as though I was being strangled by a python, squeezing all dreams, hopes and ambitions out of me. I knew I wasn't going to make it. I dropped onto my butt, wriggling out of the safety Bergen. 'This was it,' I thought. I was done. Everything inside me was screaming to quit.

Suddenly, a voice behind me said, 'H lad, you're smashing this.'

The Yorkshire accent brought me back to life. You don't hear much talking on the Thirty-miler and, oddly enough, it can seem very lonely. Despite the fact that we were in eight-man troops, it was still an intensely individual effort.

An arm came around me. 'Good fucking lad!'

I looked straight into a smiling face. It was Matt Arundel, the section commander in training and one of the top recruits. He was as tough as old boots and his fitness was absolutely crazy. He could do anything – run, swim, lift, yomp. He was a total, awesome machine.

'Fuck me, mate, we only have ten miles to push. Piece of piss. I have walked me bloody dog further.'

He slapped me once more on the back. 'You're on fire. Now get that kit back on. Let's get this done, lad.'

His words had the most profound effect imaginable. An instant jolt of hope and energy surged through me. I have never told him how much I owe him for those words of encouragement. That debt is unpayable.

Somehow I got that heavy pack onto my back. Matt smiled. 'Fuck me, you look like a turtle.'

We both laughed, and he said something I will never forget. 'All that's stopping you now, H, is you.'

He then turned and shouted to the rest of the section to come over. 'Right, lads, this is the last stretch. We're making good time, but we have to move at the same pace to make up the ten minutes we rested. So let's fucking go, because I'm bored of being a recruit!'

With that, he about-turned and trudged his way up to the next peak. We followed him, completely revitalised.

Even at that young age, I knew something special was being forged like steel inside me. Something deep within – a refusal to surrender. Matt had provided the crucial spark of inspiration. The rest was up to me.

The final few miles include a yomp across a large dam, then once again up a never-ending hill. But this time Dartmoor, against all odds, suddenly turned almost benign, as if relenting, acknowledging that we, mere mortals, had given it our best shot. The rain stopped, and all that remained of the howling wind was a light breeze stirring the straps that hung off our kit.

A recruit has to complete the dreaded Thirty-miler in under eight hours, or they fail. But it's also essential to finish in style. No one wants to stagger, gasping, across the line and fall flat on their arse. The Commander of the Royal Marines would be there to greet us, so we stopped about a mile from the end to make ourselves presentable. Rumour had it that the first question he would ask was, 'Could you turn around and do it again?'

Hopefully, it was just a question, although from what I know now, every single one of us would have done that if it came down to the wire.

Our troop double-timed into Shaugh Prior Bridge. It was over. All the shit, the pain, brutal gym sessions, freezing early-morning runs, and the endless physical and mental torment had paid off. I was a Royal Marine Commando.

I stood at attention with the rest of the Troop. Tired and in pain, we waited to be addressed.

Finally, it came. From the Commander himself. 'Gentlemen,' he said. 'This is a green beret.' He held one up.

There was silence, although I knew everyone there wanted to whoop for joy and dance a jig. But as I said, you don't do that in the Royal Marines.

'It's £12.99 at Luis Bernard's, if that's where you choose to shop.' Luis Bernard's was the on-camp shop and tailor.

'But once you have earned this, as you have, no one can take it away. It's priceless.'

We knew, all right. But no one moved. This was the top dog of the Marines talking to us as if we were his best friends.

'This beret, gentlemen, is your ticket to the party. You are now a part of the world's most elite club. You will deploy together. You will go to hell and back for one another, because we are a family. And what's more, you shall make friends for an eternity.'

Those were powerful words. I was already emotional from the drama my foot had caused me. But by then the pain was drowned by jubilation. If my foot had fallen off at that moment, I couldn't have cared less.

The Commander then walked through the ranks, personally issuing each person their beret.

'Marine Collins, congratulations.'

It was the first time I had been called a Marine – a moment I will forever cherish. I was seventeen years and eight months old.

We were carted back to Okehampton Battle Camp in a four-ton troop transport truck. I sat on the floor at the back. One of the lads handed out cigars and we smoked them as the vehicle chugged along the road. One by one, the guys fell asleep. Looking around, I saw that almost all were dead to the world, some snoring their heads off. One wasn't, and he was looking straight at me. It was Matt, the superhuman recruit who had spoken the magic words that galvanised me off my metaphorical deathbed and got me yomping again. He smiled and whispered, 'Dog-walking distance.'

I nodded, forever grateful to this amazing guy who took the time and the humanity to spur on an injured teenager to the ultimate goal. People like that are gold dust.

Once we got back to CTCRM, I headed to the medical centre for X-rays. This time I was a Marine and not a recruit. I sat there, idly flicking through a car magazine as I waited for the results.

'Marine Collins?'

'Yup.'

'You'll need to have crutches. You have stress fractures in two places from the original break.'

4

The Units

After completing training, we were sent to a unit of our choice. For me and Matt, the inspirational superhuman, this was Comacchio Group, tasked with protecting the UK's nuclear warheads.

HM Naval Base Clyde on Faslane Bay was specifically chosen to be a nuclear submarine port, as the weather is always awful in that part of western Scotland. Consequently, it's almost impossible for hostile countries to take satellite imagery through the thick banks of cloud and mist. So I knew this was no tropical paradise, but one of the reasons I picked it was because the base has an abundance of gyms, pools, saunas and steam rooms. I had become a big fan of working out, after realising the benefits of physical training in helping me to reach seemingly impossible goals.

Comacchio, now called 43 Commando Fleet Protection Group, operated on a three-weeks-on, one-week-off rotation and the downtime was a huge bonus, giving Commandos the opportunity to do other things, including pursuing the ladies in our lives. After being locked away in an all-male environment, I was certainly ready for that.

Although Matt and I were in the same squadron, we were in different teams, but we shared the same shift rotation. This led to some wild times, such as riding 250 miles to Leeds on the back of his Suzuki Bandit 600 for a piss-up and almost smashing into the back of a lorry at full speed. Matt slammed on the brakes so hard that I felt the back wheel lifting off the ground like a kite. We were very lucky. Also, it was November, and all I had on was a flimsy fleece and T-shirt. I have never been so cold in all my life.

I settled into the Comacchio routine nicely and became mates with Big Rob, a monster of a man who played for the Royal Marine's rugby team and, at the time, was training for a special forces selection course. Most Marines tend to go to the Special Boat Service (SBS) due to the similarity of the amphibious roles, but Rob wanted to try out for the Special Air Service (SAS), which is land-based, but also has a boat squadron.

Rob was down for summer selection, so he took me under his wing and I slotted into his gruelling training regime – hard yomps around the perimeter of the camp, swimming in the pool and hitting the gym in the evenings to stay strong. Like me, Rob had chosen the Comacchio Group for the workout facilities. He could train three times a day with no distractions, as the job mainly involved being on standby in case anti-nuclear protestors arrived at the site or attempted to breach the camp. As I was still fit from passing out of CTC, I trained day in and day out with Rob, and put on a bit of muscle, as I was both eating and exercising like a racehorse.

Rob then left for the SAS selection and Matt was transferred to 45 Commando, so after six months of boredom being stuck behind a fence, I applied to get a specialist qualification in driving. At that stage I didn't have a licence and envied the lads who owned fast cars and could go wherever they wanted to on leave or pop off to chat up the local girls. I had to rely on other people for lifts, such as Matt and his Suzuki Bandit – a sometimes nerve-racking experience.

The Marines sent me on the specialist eight-week course and, at the end of it, I passed an assortment of tests: motorbike, car, HGV, trailer and coach. I was drafted immediately after into the Commando Logistic Regiment (CLR) and ended up being sent to Devon. However, the camp there was full, so they couldn't provide on-site accommodation. Instead, thanks to some insanely good luck, I was given the keys to a house in the married quarters off camp.

For me, this was heaven. I had just bought my first car and instead of living in a tiny box room, I had a house to myself. The perks of being a driver were also fantastic. We had weekends off, with Wednesday a half-day to go rock-climbing, and on Fridays work finished early so the lads could go home if they wanted. Home in Lincoln had zero appeal for me, especially as in Devon I could come and go as I pleased. I became what's known as a 'camp orphan' – someone who never goes home. But at the age of twenty, with my own car and house, no camp rules or regulations, no security guards at the gates and money in my pocket, I felt like a king.

Weekends were all about drinking and women. The drinking was the easy part; women proved to be a little trickier. I would quickly hop from one girl to the next and while this may sound great, being out on the pull every weekend meant that I quickly ran out of options. In a small town near a massive Marine base, the girls soon knew we were after only one thing. So the smart ones never gave us a chance and the others ... well, they always regretted it. Local girls started avoiding me.

Consequently, we started going away for weekends where there were no Marines. Newquay, Torquay, Bristol and Weston-super-Mare were favourite spots, which meant that, in worst-case scenarios, we would have to book hotel rooms or sleep in our cars. It was seen as a game, and the more confident among us would rather rely on charm and attempt to 'trap' – the Marine term for going home with a girl – rather than waste money on a room. Armed only with the confidence of youth, I was quick to roll the dice and gambled on never booking into a hotel or packing a sleeping bag in the car. We invariably started hitting the bars as a large group, which would quickly dwindle as the lads sought out targets and a bed for the night. Some guys had tactics, while others relied on sheer instinct. My thing was divorcees, as I found that girls of my age usually bored me. Older women had far better conversational skills and I enjoyed their company more. It always takes two to tango and I like to think they got as much from it as I did – it wasn't always about finding somewhere to sleep for the night. Maybe I'm trying to justify the irresponsibility of it all, but I enjoyed that wild-oats period of my life immensely. I was carefree and very much living the dream. However, my view on women was that of a cocky twenty-something and I wasn't the best human being, causing quite a lot of upset back then.

Luckily, the women of the South West would soon be spared my antics, as I was about to be sent to the Balkans. It was 2000, the start of the new millennium, and our unit had been deployed on a NATO-led operation in Kosovo to support the multinational brigade out there. This was my first op (Marine lingo for an operational tour) and Kosovo was a bleak, war-torn country with little going for it. The entire Balkan region was in chaos, as the former Serbian leader Slobodan Milošević had committed horrific atrocities against Croats, Muslim Bosnians and Kosovan Albanians, only stopped by seventy-eight days of intense NATO bombing. The devastation had to be seen to be believed.

We arrived at the back end of summer and were warned about the coming brutal winter, so I was not looking forward to it. However, I soon settled into the odd routine of an op, and it was a lot of work getting the ruined country up and running again. We mainly patrolled vulnerable neighbourhoods to keep residents safe, as well as providing logistical and humanitarian aid, such as delivering food and water.

Life was fairly pleasant and the gym more than adequate. We even had a boxing competition between the two Marine units in the country. With the warmer weather ending, the temperatures began to drop and that seemed only to add to the harshness of the place. Our camp backed onto a bombed-out factory, with large bits of metal sticking out all over the place, and when the snow fell in thigh-deep drifts, it was little more than a drab, frozen wasteland.

My first Christmas overseas was a mad affair, as the Marines don't do things by halves. Officers served the junior ranks turkey and all the trimmings, as was naval tradition, and beer, wine and rum flowed freely. Stuffed with food, booze and good cheer, many lads only made it to the early evening before passing out. I retired at about eight that night, just as the more hardcore in my unit got started on a case of port.

Kosovo has been through more misery, bloodshed and destruction than most other European countries, and it was quite sobering for me, barely an adult, to see all this first-hand. Some of the stories of brutality I heard were shocking beyond belief – men herded into groups and shot, women and kids murdered, multiple gang rapes, forced evictions – and we saw several mass graves where victims of genocide had simply been dumped and covered over. I could smell the decomposing bodies and some nearby trees still had bloodstains on them.

As spring arrived, our unit started winding down our work and it was time to go home. I had saved a lot of money during my time away and was keen to buy a new car, which I hoped might impress some ladies when I returned to my old ways in Devon.

July was the annual summer leave month in the Marines, which usually meant three weeks' holiday. I decided to stay in Devon, as I loved my set-up at the house. After all the destruction we had seen in Kosovo, the rest of the guys in my unit just wanted to go home, so the town was empty. I found a wingman who had also decided to stay in Devon, and for three long, lazy weeks we spent our time loafing on the beach, drinking and attempting, very feebly, to stay out of trouble.

By August, with alcohol levels at an all-time high, I reluctantly went back to work. On arrival, we were told we would be gearing up for some sort of exercise, but half-drunk and half-hungover, I didn't pay much attention at the briefing. Talking to some of the lads afterwards, it turned out that the exercise in question was a major joint operation between the UK and Oman, which would be taking place in the Arabian Desert that October. Called Exercise Saif Sareea Two, it would involve all aspects of military hardware working in conjunction with one another.

To prepare for this, we would be doing some training on Dartmoor, so once again I found myself at Okehampton Battle Camp, the starting point of the Thirty-miler. The main difference was that this time it was in September and we were mobilising for the desert, not a yomp in freezing conditions.

Our training programme consisted of firing heavy weapons, drilling in our kit and equipment, making sure combat drills were nailed down, and familiarising ourselves with equipment that we didn't use that often. My favourite was the .50-calibre, a beast of a machine gun that could fire clean through a wall and really mess up your day. We also practised with mortars and anti-tank weapon systems, and it was an exciting shooting package, unlike the standard firing-range days with only SA80 assault rifles. The unit was soon fully prepped and ready for the marathon desert exercise.

Then, everything changed. When we returned to Okehampton in the afternoon, after having been on the moor all day, the camp was unusually quiet. Eerily so. Lads were normally outside, playing football or smoking, but it was like a ghost town. Our four-ton trucks parked outside the Navy, Army and Air Force Institute (NAAFI) room and when I looked through the window, the entire unit appeared to be crammed inside.

'What are they all in there for?' I asked.

No one had an answer.

As we jumped off the back of the trucks, the door of the NAAFI room swung open. It was the Troop boss.

'Get in here, lads,' he shouted.

He didn't seem happy and that usually meant one of us had messed up massively. However, we had been shooting all day and, as far as I knew, it had gone well. We didn't kill anyone, which was always good when practising with live ammunition. We squeezed inside, fighting our way around lads in other companies, to get to the TV at the front of the room.

That's when I saw it. I watched the screen in awe as huge fireballs exploded. Two commercial airliners had flown into New York's Twin Towers.

The date was 11 September 2001.

5

Afghanistan – Round One

The room was silent. We hung on every word the newsreader uttered, while a small inset on the right-hand side of the screen continued to show live feeds of the Twin Towers, flames and smoke billowing. It didn't seem real and I could not take my eyes off the images.

It was clear this was a deliberate attack. But by whom? And why? All sorts of theories flew around the room, ranging from American IRA sympathisers to Saddam Hussein. All I knew was that whoever had done this would be hunted down with a vengeance by the Americans – and, by default, we would go along for the ride. After all, when the Americans sneezed, it was the Brits who caught the cold. Particularly the Brit military.

The Troop boss had a sister who worked in one of the Twin Towers, so I understood his agitation when he called us into the NAAFI room. Every few minutes he frantically dialled New York, until he eventually got through and discovered that, by extreme good fortune, she had not been working that morning.

The events of 9/11 changed the world we live in forever. I was no exception. My life was about to be turned upside down, intricately linked for more than two decades to what the then American President George W. Bush termed 'the war on terror'.

Despite the looming near certainty of conflict, the Oman desert exercise was still on. However, we were then told to pack for a 'long period' and possibly also an operation, but given no time frame. That's like trying to dress in the dark with one hand – what gear should we take, and what should we leave behind? What was essential? And what was not? We were never told what to take on any exercise. However, we were expected to be on the ball with everything: kit, weapons, knowledge of the country we were going to – the lot. As long as you were well organised, in the Marines you were left to your own devices. If not ... well, someone would pay you a little visit to let you know in no uncertain terms. I was already packed for the Oman exercise, but I had to be prepared for a potential op as well. My Bergen was rammed beyond capacity.

We hit Oman in style, with a full-scale beach assault at the mouth of the Persian Gulf, landing on a scorching, windy afternoon. Our vehicles were open Land Rovers, so we got blasted with hot air and stinging beach sand that got into every orifice, as well as into our kit and weapons. We then went deep into the desert to rehearse simulated battle scenes with Omani tank and infantry divisions, as well as a few in-house scenarios with our units. But although this was one of the British military's biggest peacetime exercises, the talk was of only one thing: war. The follow-up to 9/11.

On 10 September 2001, few people in the West had heard of either Osama bin Laden, al Qaeda or the Taliban.

Twenty-four hours later, they were household names, although I doubt that many could actually pinpoint Afghanistan on a map. But what everyone in the Marines did know was that from Oman it is a mere puddle jump to Afghanistan, less than three hours by plane. So if we were to hunt the mastermind of 9/11, surely we would go in straight from there? What was the delay?

As was the norm in the Marines, rumours flew at supersonic speeds, and since we were out in the desert and bored, the grapevine took on a life of its own: some of it true, some of it hearsay and the rest bullshit. One minute we were about to deploy, the next we weren't. The only thing certain was that no Marine wanted to miss being the first on the ground in the looming war. My biggest fear was that my unit, Commando Logistic Regiment, would not be involved in the op and I kept every finger crossed, hoping not to be left behind.

I soon got my wish. A couple of days later we learnt that we were indeed being sent to Afghanistan as an advance force, and would soon be flying into Bagram Airfield. Although it was still unsecured, talk was that the SBS were going in on a top-secret mission to take it. Or so we heard. One lad in our unit had a brother serving with the SBS in Poole, but as he refused to say anything, we had no idea.

Afghanistan is a landlocked country that for centuries has been referred to as the graveyard of empires: many have come to conquer and all have failed. The terrain is savage, as are the searing summers and freezing winters, and the Afghans are among the toughest people I have ever met. They have suffered beyond any comprehension and yet are never defeated. Even the might of the Soviet

Union could not prevail against them. In 1989, after a ten-year occupation, the Russians ignominiously withdrew, allowing the Taliban, adherents to the most austere form of Islam in the world, to take over and rule in line with their precepts. This led to the 9/11 attacks, and Osama bin Laden, who had been in Afghanistan since 1996, was the world's most wanted man.

He was somewhere down there below, as I looked out of the window of the C-130 Hercules. I could feel the plane dropping in increments as it carved its way between the high peaks of the Hindu Kush mountains. For once, the Royal Marines were landing in daylight, making it easier. I have always hated night moves that involve aircraft.

We began a tactical corkscrew descent: the plane banking sharply to avoid gunfire, before levelling out onto the rough tarmac of Bagram Airfield. As we taxied down the runway, I could see wrecks of Russian MiGs pushed off to the side – decaying and rusting monuments to Soviet failure.

The loadmaster stepped over us and punched the button to drop the back ramp, and in poured fiercely bright daylight, forcing us to squint. We disembarked, and a guy wearing US combat trousers and a white T-shirt greeted us, looking as though he had just got out of bed. We followed him off the runway and over to an area with tents and some quad bikes, and he pointed out where each unit would be staying.

BOOM! A massive explosion only a few hundred yards away caused everyone to jump and duck for cover.

'That's the EOD (Explosive Ordnance Disposal) teams,' said white T-shirt man. 'The airfield is toppers with mines.

Someone stood on one yesterday so, for fuck's sake, stay on the concrete!'

I took note of his instruction.

We moved over to our accommodation. It was a full-on shanty-town set-up: rows of tents at the side of the airfield. The SBS lads were staying in a large hangar and the compound further down was 45 Commando's camp.

For the first few nights we did security on the airfield, sleeping at the side of the runway in shifts of two hours on and two hours off. In my spare time I went over to the 45 Commando tents. They were a great bunch of lads and I knew a few from my time in Scotland, but also the food there was far better. On one occasion, as I walked into their camp, I passed three lads working on a WMIK (Weapons Mount Installation Kit), a modified Land Rover with reinforced chassis and mountings on roll bars to run heavy calibre weapons.

'H, how the fuck does this work?' asked one.

They were trying to figure out the vehicle's differential lock, and I started explaining how to operate it. To be fair, I only knew that thanks to the specialist driving course I had been on some months beforehand. This kit was fairly new to the units, so a lot of the lads were feeling their way around.

After a few minutes, I was interrupted. 'Mate, do you want to drive for us on this op? We need two lads for our support vehicles.'

I answered almost before he had finished asking. 'Definitely, dude!' I couldn't believe my luck. My unit, the CLR, was staying behind in Bagram and there was no way I wanted to do that. I would instead be on loan to 45 Commando and go on the actual frontline operation.

I knew exactly who to pick as the second driver and took off to ask him if he also wanted to come along. He jumped at the chance.

The op would be waged from Khost, the capital city of Loya Paktia, a strategic region in the south-east of the country where the Taliban had grouped. We would have to drive over the Khyber Pass, a treacherous mountain passage through the notoriously lawless Afghanistan–Pakistan border area. It's one hell of a drive up narrow, swooping, potholed roads, taking about six hours. But for us, with so many vehicles in the convoy, it would be a two-day journey. The roads were badly damaged, and in some places, little more than crumbling tracks.

We rose early, getting our vehicles ready for the gruelling expedition. Bagram Airfield is surrounded by rugged mountains and sunrises are especially beautiful, as the morning light bathes the peaks in stunning shades of pink and purple. Against the backdrop of the jagged, snow-capped Hindu Kush, it was simply breathtaking. The WMIKs, covered in camouflage nets and .50-cal machine guns bristling on roll bars, contrasted starkly with the panoramic scenery but told the true story of the land: Afghanistan was as deadly as it was beautiful.

Before leaving, we were issued with all the latest Gucci kit, a Marine term meaning good equipment. In this case it was state-of-the-art Garmin GPS and satellite phones, but what we craved above all else were American food rations, or MREs (Meals Ready to Eat). As with most things American, MREs were bigger and better than our food packs, which were positively vile. Even when starving on Dartmoor, it was a struggle getting Brit army rations down our throats and they usually made me feel slightly sick. Conversely, our cousins from across the pond had decent meals, such as beef and noodles or burritos, and my personal favourite, chilli con carne. In addition to the nice meals were Skittles, Hershey bars, peanut butter, crackers and a small bottle of Tabasco sauce, which sparked my love of spicy foods. I doused everything in Tabasco. We threw a load of the MRE boxes onto our WMIK as we loaded up the ammo.

The convoy of mobile killing platforms and support vehicles started moving mid-morning, engines rumbling into life as we drove off. Carcasses of rusting Russian tanks and trucks sprawled on both sides of the road – stark evidence of how bearded men wearing flip-flops and living in caves had defeated one of the biggest global superpowers the world has known.

'Graveyard of empires,' I whispered to myself as we passed another piece of junked hardware. Alarmingly, even after all those years, much of the ordnance scattered in the countryside was still active – mainly anti-personnel mines that had been dropped by Soviet aircraft. These plastic devices looked like butterflies, but that is where all similarity ended. They were designed to take a foot off whoever was unfortunate enough to step on one. Or in the saddest cases, the hand of a poor curious child picking up one as a toy.

I can only describe my first views of Afghanistan as a scene from *Star Wars*. It genuinely felt as if I had landed on some distant planet. Everything was alien: the clothes, the traditional mud houses, the bustling markets with buyers and sellers haggling at full volume, and the open suspicion of people just stopping in the middle of whatever they were doing to stare at us. The women were covered from head to toe in blue burkas, with a thin strip of mesh at the front for their eyes, reminding me of the Sand People from the George Lucas movies. Most women have a good life in the West. Not here. They always seemed nervous and uncomfortable, and I later heard many stories of their systemic oppression by the Taliban.

The road started climbing, indicating that we were approaching the famed Khyber Pass. The higher it got, the more treacherous it became and the greater the likelihood of deep snow. My WMIK was cumbersome and dangerously unbalanced with the heavy weapons systems bolted on. To add to that, ammunition in metal containers hung off the sides, banging into rocks every time I turned up a narrow path.

It was December, the start of winter, yet the sun was actually quite pleasant. So much so that it reminded me of summer road trips to Newquay and Plymouth, which was bizarre because, since 9/11, carefree times seemed to be memories from another life. But now, driving high into the mountains of the Afghanistan and Pakistan border, it all came back: the nights chasing girls, drinking beer and watching sunsets on the beaches of Devon.

Not for long. The terrain got more and more challenging, and I was jolted back to reality. I could see why the Russians had never defeated the Afghan mujahidin. These hardy mountain men knew every nook and cranny of this harsh land. The mountains offered perfect vantage spots for ambushes and the Russians didn't stand a chance when caught in deadly cross-fire on narrow passes surrounded by steep crags and cliffs. Scores of burnt-out tanks and trucks bore mute testimony to that.

We drove in the dark, using night-vision CWS (Common Weapon Sights) or Cyalume sticks to guide us through the star-studded night.

Finally, as dawn broke, we reached Khost.

6

Operation Snipe

The convoy stopped at a large, open compound, which was to be our main base for the next few days. From there, we would hunt down the enemy, and maybe even find the mastermind himself: Osama bin Laden.

Just as it had the morning before at Bagram, the rising sun blessed us with its beauty, bursting up over the mountains that spread eastwards in craggy ridges to Pakistan, about twenty-five miles away. Despite the pastoral scene, our camp wasn't exactly a haven of tranquillity, as a bunch of Northern Alliance guys were screaming about in an old T72 Russian tank that had seen better days. In fact, it was a miracle it was still working. There was not much we could do about the racket it made, as the Northern Alliance was our ally, having fought against the Taliban in the civil war that erupted after the Soviets left. However, their ill-disciplined, extremely disorganised – and, in some cases, unhinged – fighters did not inspire much confidence. But it was not my place to judge. I had just turned up and was still mentally connecting the chaotic jigsaw pieces of Afghanistan.

We set up a position out of the sun, along a ditch that afforded cover should anyone decide to say 'hello' with an AK-47 or some IDF – the term for indirect fire, such as mortars or rockets. I popped in my Minidisc player headphones, rested my head on a roll mat and shut my eyes.

Before I knew it, I was shaken awake by one of X-Ray Company's HMG gunners.

'H, looks like we're on, mate.'

HMG stands for Heavy Machine Gun, and they are beasts. So are the guys who operate them – big, powerful men handling big kit, and you don't want to mess with them. They have a rock-solid reputation in the Corps, and they were way more experienced and capable than I was as a young Marine. I followed him, still half-asleep, into an olive-green tent. The headsheds (officers in charge) were gathered at the far side of a table laden with maps, markers and some model cars. This was our makeshift operations centre, hastily put together with the minimum of resources.

'Gentlemen,' said one, 'first of all welcome to Khost.'

I had not seen him before but I figured he was some sort of intelligence officer. Using the maps, he carefully explained when and where we would go in, and exactly what we had to achieve to complete the mission.

'The name for it is Operation Snipe,' he said.

I could feel the excitement in the air as the lads stirred. We would be hot-pursuing the Taliban into their mountain hideouts, driving them out of the country into Pakistan, as well as disposing of any heavy-hitting hardware and ordnance we found along the way. It was one of the best ops that the units had been on for a long time and everyone was listening with full concentration – a rare thing for young Marines.

The headsheds ended the meeting by telling us that 45 Commando, my temporary unit, would be moving out the next day. No big deal, as most guys were already packed. All I needed was a good night's sleep after driving through the previous night. However, the two clowns in the T72 seemed not to have tired of thrashing the tank up and down our position, which quashed any hope of me resting.

Early mornings have never been my thing, even though I have spent most of my life getting up at crazy times. I still hate them, particularly after a noisy night in a ditch, with stones digging into my body. But despite being pissed off and the lack of sleep, I was eager to get going on my first proper operation. Kosovo had been good, but Op Snipe was on another combat level altogether. For a start, us chasing the Taliban was all over the news and people were out for revenge after 9/11. This was a high-profile mission.

We were told to expect resistance as we closed in on the Pakistan border. There seemed little doubt that my first contact would be in the coming days, but I don't remember feeling nervous. It was more of a giddy excitement. The real thing was happening. We had the kit, the men, the firepower and the motivation. All we needed was a dance partner, and he wasn't far away.

The Commando moved off, weaving out onto Khost's dusty, hard-packed roads and up into the winding mountain passes. There was one hairpin turn after the next, and the overloaded vehicles in the long convoy lurched dangerously from one side to the other. What made it more treacherous was the fact that local truck drivers, bombing down the pass, would take the sharp corners at Mach 1 speeds while we were on the edge of sheer cliffs, missing us by inches. Jesus, that made me nervous. I enjoy a challenge, but suicide wasn't up there for me.

The skies started turning grey and the heavens opened, making driving even harder. The long, snaking convoy of heavy-hitting hardware halted and the boss came running down the line, stopping briefly at each vehicle. He eventually reached me.

'This is us for the night, lads. Intel says we have Taliban up ahead and we are waiting for a green light. Set up into all-round defence, and put the .50-cals on the high ground.'

We were on a ridgeline, separated by a large wadi from the other side of the hill, where a local had said there were Taliban fighters. We kept as low a profile as possible, trying not to spook anyone, although if the enemy were there, the locals would have told them about us. It would not be a big secret.

Nightfall came and the company shook out into a triangle formation, protected on all sides with heavy machine guns on each corner. I was tasked with manning the point nearest the wadi, as the HMG lads had incorporated me into their sentry rota, doing one hour on, two off. It was always lonely on guard duty, and while sitting behind the .50-cal, scanning about with my night-vision sights, my mind would wander off on all sorts of pointless shit. But the hour passed quickly and I went back to my roll mat in the sand, closing my eyes and trying to get to sleep fast, as the two hours I had to rest were already ticking down.

CRACK, CRACK, CRACK!

The unmistakable staccato of AK-47s woke me with a start. I quickly grabbed my kit and moved towards the noise. The lad on the .50-cal was already swinging it towards the direction of the gunfire, as another salvo slammed into our position.

The gunmen seemed to be shooting from across the wadi, and as the next volley of shots cracked past us, I saw them. They were in a car, whose silhouette stood out against the mountainside. The mortar lads quickly shot off two illumination rounds, lighting up the sky and pinpointing our target. We could see them clearly then: four men in what appeared to be a blue or purple car.

The .50-cal gunner fired, and we followed suit with our SA80 rifles. The car screamed off and out of view. We stopped firing and the mortar guys put up more illume that lit up a small village towards which the car appeared to be heading.

'Fuck me, they have to be hit,' said one of the guys next to me.

I nodded. We were all confident our rounds had found the target, and Ears – the HMG lad – had definitely scored a direct hit. We crossed the wadi and into the village, searching compounds and waking up every resident in the process. Eventually, they showed us where the gunmen were – or at least where they had been. We found a hastily parked car riddled with bullet holes and every window shattered. Around the vehicle were large pools of blood, glistening wetly in the starlight. One of the locals then told us that the gunmen were four brothers who, thinking we were Taliban, had wanted to scare us off and shot into our position. Bad idea. They had all been hit and were being taken in another car to a hospital in the next town.

Well … that was the locals' version of what had happened. But few, if any, of us bought it. It was stretching it a bit to believe that the brothers had mistaken a massive convoy driven by foreigners through populated regions for cunningly disguised Taliban fighters. We had no doubt the 'brothers' were in fact Taliban, but couldn't prove it. One thing for sure was that those dudes had been seriously injured and had lost a lot of blood. I never found out if they survived. The moral of the story is: don't shoot at what you cannot see.

We continued chasing the Taliban towards Pakistan. The days became weeks, and we all started looking more and more shabby. The once smart, well-groomed young Marines were looking like a bunch of homeless people with beards, long hair, hollow eyes and dark suntans. It never bothered me not showering or wearing deodorant, as our training prepared us for weeks of living rough, but we were a sight.

One of the positions we occupied had a well-concealed flat rock and the boss decided to place a L96A1 sniper rifle on it with some NVDs (night vision devices). It was a great position: the arcs of fire were fantastic and the view went on forever, so no one could sneak up unawares. It only had one flaw – the rock itself. I was young and didn't know that sitting on a cold hard surface for hours on end would have an impact on my butt. I become more and more uncomfortable, and eventually snapped and went to see the company medic.

'Hey Andy, what's going on with my arse?'

A volley of quick-witted and sharp-tongued answers followed, but for once I wasn't in the mood for banter. I needed to get rid of this pain in my arse, so to speak.

'Mate, it's haemorrhoids,' said Andy. 'You'll live. A helo is coming with a resupply the day after tomorrow and I'll get some cream ordered.'

I have to say those were the longest two days of my life, as I'm rubbish if I'm not functioning at a hundred per cent. It finally arrived – that big, beautiful twin-rotored machine, dropping out of the sky to help my poor bottom. I'm sure it served other purposes, too, but for me that was all that mattered.

Someone chucked me the tube of cream, along with a smart comment, but by then I had probably heard them all. I wandered off to find a private spot, whipped down my combat trousers and boxers, squatted and proceeded to lather up my bits.

'Oh my God, that's good!'

I have a habit of talking to myself. But to my surprise, that time I wasn't. On the ridgeline were about twelve somewhat bemused Afghans who had come out of nowhere and were standing above me on a large rock. I was probably the first Westerner they had ever seen – and certainly the first with his pants down and rubbing cream into his butt. I can't even imagine what they were thinking.

With my work done and my audience still watching, I pulled my pants up, gave them a wave and headed off. It was only afterwards that I realised I had waved to the locals with the same hand that I had used to apply the butt cream – probably the most insulting thing possible to do to an Afghan. One thing is for sure, I was never going to get a job in international relations.

During the war against the Soviets, Taliban fighters had dug a vast network of caves and depots that stored everything from RPGs (rocket-propelled grenades) and explosives to surface-to-air missile launchers and anti-tank systems. Each den was an Aladdin's cave of war-fighting kit, and we were instructed to find them.

One cave we unearthed was so massive that I walked for an hour in it before turning round and heading back. It had everything: carpets, chairs, beds and fully equipped armouries with enough gear piled up against the walls to start World War Three. The engineers decided to demolish it and our task was to cover them as they rigged it up. It was so huge that it took them three days to place all the explosives.

Once done, we all pulled back and the order was given. It was the largest explosion since the Second World War and I was told it changed the topography so much that they had to redraw maps of the area.

Operation Snipe lasted about four-and-a-half months, and it was now mission accomplished. The lads had performed exceptionally well, and most of the Taliban had bolted across the border into Pakistan, where we could no longer chase them. We still came across the odd body of a fleeing fighter killed by American air strikes, but as an army, they had been utterly routed. We had done what we had set out to do and helped to stabilise the country.

For the moment, anyway. Little did I know then, but twenty years later that piece of history would repeat itself. Only, this time I would be the one running.

Stinking, underweight, hungry and filthy after having lived like hoboes for so long, 45 Commando returned to Bagram. We barely recognised the place. Rather than being just an airfield, it had been transformed into a bustling small town, as the Americans had thrown up all sorts of buildings. I moved back to the Commando Logistic

Regiment, and for the next few days revelled in taking solar-heated showers, eating decent food and enjoying American entertainment.

But then it was time to go home. Trouble, it seemed, was brewing elsewhere.

7

Somerset and Saddam

Soon after returning from Afghanistan, I was transferred from Commando Logistic Regiment to 40 Commando. My new base was in Taunton, Somerset, where the cider was strong and the ladies plentiful. In other words, my kind of town.

I had also hooked up with an older woman called Jane. She mothered me a bit and I think she felt more sympathy for me than anything else. We were polar opposites who somehow had got together, but I really liked her company. She was interesting to listen to and quite a spiritual person; she had a beautiful big old house with a large entrance and grand piano. She loved to cook, and would regularly get dressed up to the nines and invite me over for gourmet meals. I never seemed to be dressed for the occasion, as I either wore a gym kit or jeans and T-shirt, but she never mentioned it and was always happy to see me, despite my lack of effort.

We had been dating for about six months and things were starting to spice up, with Christmas not far off. Then, just before Christmas leave, 40 Commando headshed issued a 'clear lower deck' command, which meant that every serving and attached member of the unit had to attend. We figured it would be the usual festive season 'don't drink and drive or beat up family members' pep talk, but in the hall were some pretty big players. The kind of top brass who don't turn up for trivia briefings.

The commanding officer stood. 'Gentlemen, there is no good way to say this, so I will just say it. Your leave is getting cut short.'

Heavy sighs and 'fucking hells' muttered through gritted teeth filled the air.

'You will have Christmas and New Year off. But on New Year's Day, 12 p.m., I want every member of 40 Commando back here.'

I knew what was coming next. We had all being watching the news reports that Saddam Hussein was messing inspectors around regarding his supposed 'weapons of mass destruction' stash. He refused to hand them over and was hampering UN officials at every turn. The UK and the US were getting increasingly impatient, and had started issuing ultimatums.

The CO continued, 'We will be sailing on 2 January from Marchwood Military Port. Pack desert gear. If it goes sideways in Iraq, you, gentlemen, will be the tip of the spear. It's a good op so do not screw it up by getting locked up over Christmas. Enjoy yourselves but be ready to operate come January.'

Ah, there it was, the 'don't bring the Marines into disrepute' duty lecture. I knew he would slot that in somewhere.

Mixed emotions swept through the unit. Many, like me, were excited, but others were angry about leave being cut short. Some weren't happy about either the op or the cancelled holiday.

'It will never happen,' said one of the lads. 'We'll just have a shitty New Year, come back here, get messed about at sea and then have to sail all the way back. Waste of fucking time, if you ask me!'

I suppose you can't please everyone, but I couldn't understand this negative headspace. Personally, I was in a good place. If it went ahead, I would be on my third op and I was barely twenty-two years old. If it got binned, then I would have Jane and her big house to go back to. So it was a win-win.

But even so, in my opinion, if you didn't want to go on ops, there was not much point in being part of a rapid-reaction regiment.

Later that evening I broke the news to Jane and she was gutted. She really wanted to be with me over the festive season. That worried me, and I started thinking that perhaps I had been spending too much time with her. She had grown too accustomed to seeing me most days. I had to admit to myself – if not to her – that going on another op appealed to me far more than staying put in suburban bliss.

I was on duty on Christmas Day, but Jane invited me over to her place for New Year's Eve. I grabbed a bottle of red wine from the supermarket near camp and wound my way to her house along Taunton's country roads. With money saved in Afghanistan, I had bought myself a Toyota Celica VVT-i, which looked sexy as hell in shotgun silver and handled the back roads like it was on rails.

I pulled up outside Jane's place with a heavy heart. I knew I should break up with her before this op but didn't have the guts. Messing girls about was one thing, but she was older and had already been treated badly by her ex-husband. She was in a delicate place, hence the spiritual stuff. In the long run I was probably just a phase for her anyway, but even so, I could not bring myself to add to her misery. I got out of the car and decided to have a bloody good blowout before heading off for Iraq.

Taunton to Marchwood Military Port near Southampton docks is a fair drive. With a red-wine hangover, it's horrific. I leant out the window and puked pure crimson on the road. Jane and I had done four bottles between us that night and I barely remembered a thing, so it must have been a good 'hello' to 2003.

We loaded kit and equipment throughout the night onto HMS *Ocean*, which would be our home for the next few weeks. The vessel is a specially designated Royal Marine transport ship; its prime function is to move Marines to whatever op is on the go, so one would have thought that Marines would be accorded some respect.

Not so. In fact, it was the opposite, as the navy treated us as one would a disliked family member. We could eat only after they'd eaten and could train only once they had done so. I know this got a lot of our guys' backs up, as without Marines, HMS *Ocean* would not even exist. Yet we were the unwanted guests. It was hardly conducive to brothers-in-arms camaraderie and certainly the navy's arrogant attitude didn't help.

The situation in Iraq was close to an outright declaration of war, so a fleet consisting of HMS *Ocean*, the aircraft carrier HMS *Ark Royal*, and Royal Fleet Auxiliaries *Sir Galahad* and *Sir Tristram* as support vessels sailed to Cyprus, where we were to run a full-scale weapons test of all our unit's hardware. However, we were more excited about a night out in the local *tavernas* than any training exercises. Even strict instructions that no ranks were to leave camp did not deter us in our quest for beer and women. Sneaking off the military facility became a full-time game of cat and mouse between us and the camp guards, with the latter usually winning.

In the end, we admitted defeat and left camp only to conduct long, hot shooting packages on the local ranges, using every weapon in 40 Commando's vast arsenal. The majority of us were well-versed in most of the weaponry.

From Cyprus, we headed directly for the Persian Gulf, eagerly waiting for a heads-up on what the potential op would be. We didn't have to wait long, as on 10 March a mission order came through with our specific directives. It was top secret, which meant that everything had to be shut down for maximum security. Nothing would leave the ship. No phone calls, texts, messages, emails or letters. The op had become reality. Its name will go down in history as the first battle of the 2003 invasion of Iraq: Operation Telic.

The word 'telic' actually means a 'defined action' in Greek, but as it had been planned over Christmas, the lads referred to it as 'Tell Everyone Leave is Cancelled'.

One by one, the teams were called into the ops room to be briefed. In the centre was the best model of an attack plan I have ever seen. It was a work of art, depicting a huge complex of buildings, pipes and machinery. Even before anyone spoke, I could see it was some sort of oil facility.

That was soon confirmed by the officer briefing us. 'This, gents, is the PPLS oil terminal.'

Our job was to capture the oil pipelines, metering and pumping stations – an area of one-and-a-half square miles on the strategic Al-Faw peninsula, through which ninety per cent of Iraq's oil is exported. Then we would seal off the entire peninsula – a sliver of land between Kuwait and Iran – and hold it against any counter-attack by the Iraqi Army.

In short, we would be the first troops on the ground in Iraq. We would be the guys kicking the door down. The task facing us was huge and the briefing extensive; hence, the finely detailed model before us of how the attack would unfold. We certainly had our work cut out: we were outnumbered, the Republican Guard – Saddam's elite force – was just up the road, in Basra, and there was also the very real possibility of a nuclear, chemical or biological attack.

'Fuck me, this will be messy,' said one of the commandos next to me. My sentiments exactly. At that stage everyone thought Saddam had weapons of mass destruction.

There were a lot of moving parts involved – always a problem in any operation – and soon we were having daily briefings. Basically, the battle plan for our unit was that US Navy Seals would first breach the refinery's vault-like doors, and 40 Commando would then secure the heavily guarded complex and prevent Iraqis from sabotaging or setting fire to it. That's what had happened in 1991, the first Gulf War, when oil wells set alight by retreating Iraqi forces in Kuwait caused huge ecological damage.

In preparation for the attack, nuclear, biological and chemical (NBC) preventative procedures were given top priority, followed by weapons tests off the side of the ship. Physical training was early morning and late at night, which was second nature to us. Marines train daily, regardless of how busy the day is, for the simple reason that war-fighting is brutal and we had to condition ourselves accordingly.

Helicopters were crammed into every available inch of space on HMS *Ocean*'s lower deck. A giant platform-lift manoeuvred them up and down from the top deck, and when that was operating, all the ship's lights were doused, replaced with low-visibility red light to ensure an almost total blackout. The op was gaining momentum and we were convinced that pretty soon we would get the go-ahead.

Eventually, it came. The lowest part of the ship was called 7 Golf and housed most of 40 Commando. Normally, we were a slow-morning bunch of guys, but that morning the corridor seemed a lot busier than usual. Then an NCO came out with a towel around his waist, fresh from taking a shower.

'We're a go for tonight. Boss just said. He's been in ops room since 4 a.m.'

'Nice one,' a chorus rang out in unison.

After three months at sea, the unit was champing at the bit to get off the overcrowded ship. We were bored to distraction with briefings, and living off rumours. At last something seemed about to happen.

The lad telling us we were attacking that night was right. Instructions were to gather in full kit at the helicopter ramp at 8 p.m. and with NBC respirators – gas masks in civilian terms – ready for immediate use. We stood on the platform, bathed in the dully menacing red light, with belt-fed, automatic and sniper weapons either slung across our backs or carefully placed across the toecaps of our boots. No Marine *ever* puts weapons systems on the floor. I watched the red light silently, waiting for it to start flashing, which would indicate that we were about to board the helos.

Then came the order. 'Stand down, stand down.' It was followed quickly by our reaction, 'Oh for fuck's sake!'

'Bad weather, lads. The op is delayed twenty-four hours.'

Operation Telic would have to wait until the following day. So it was back to bed and listening to music on my Minidisc player.

At 8 p.m. on 19 March 2003, we headed back to the helo ramp, again in full kit with NBC masks. The Ministry of Defence had pumped us full of all types of weird and wonderful injections, including an anthrax vaccine. Rumour had it that anthrax biological weapons were in Iraq and if past history was anything to go by, such threats had to be taken seriously. Extremely so. Saddam Hussein had used bioterrorism before, on the Kurds in Halabja during the Iraq–Iran war, and devastating scenes of civilians, including children, choking to death in the streets had shocked the world rigid. It was seriously nasty stuff and a horrible way to die.

Once again, fully laden with kit, we trudged out of the living quarters and onto the helo lift, bathed in red light. Above us, I could hear the thump-thumps of helo rotors warming up, as well as the helos putting in quick loops around the ship before landing again, ready to pick us up. We were lined up in chalks – one chalk per helo. I was in chalk 3, and in the dim light I could barely make out chalk 2 in front and chalk 4 behind me. A few lads I knew gave an excited thumbs-up in the darkness, or a set of white teeth would appear in a grin behind camouflage paint.

The radio burst into life.

'All call signs stand by, stand by.'

Operation Telic – the largest deployment of British forces since the Second World War – was about to kick off.

8

Operation Telic

The red lights flashed slowly on and off – a warning that the ramp was about to move. The hydraulics of the lift clicked in with a smooth hum, taking us the twenty or so yards up onto the flight deck.

The helo rotors were already spinning, their heat and noise unmistakable, and a warm mechanical smell of oil and metal filled my nostrils. Quickly, the chalks made their way to their designated helos, mostly British or American Chinooks. This airlift was huge and, as US Navy Seals were involved, we would be utilising the assets of both countries. My chalk's helo was an American one, painted black and piloted by two Yanks. We snaked our way across the flight deck to the rear of the bird, where we were counted in.

My fighting kit weighed a ton. So much so that I would be glad to fire off some magazines just to drop a few pounds. Water bottles also weighed me down, as we had no idea of the drinking-water situation, so best to be on the safe side.

We sat there for what felt like an eternity, impatient and restless as the rotors thumped, and still not convinced that instead of a green light we would be told to stand down – as had happened the night before.

The whine of the helo suddenly intensified. The op was a go. Helo after helo lifted off the deck of HMS *Ocean* – a giant body of men and machinery heading for Kuwait to pick up extra people on the Iraq border. From there we would have Alpha, Bravo, Charlie and Delta companies, including attached ranks and teams from 3 Commando Brigade HQ and the US Navy Seals, all rolling into action. I was in Charlie Company.

Within five minutes of landing in Kuwait, we were up and off again, for which I was grateful. My coffee intake had been high and my bladder is small, and I knew it would definitely be a while before I got the chance to take a piss.

With all the turbulence, the helos bumped around like small boats on a rough sea. Then, without warning, the loadmaster standing by the ramp frantically started looking out of the back, desperately searching for something. We had no communication with the pilot, and the engine noise was far too loud for him to hear anything we said. But we all knew that whatever the loadmaster was looking at, it wasn't good.

A few minutes ticked by and then a voice came over our personal radio network.

'Lads, a Brigade Patrol Troop helo has just gone down. We need to focus on the task at hand, so stay on form, boys.'

I had some good mates over at BPT – a crack surveillance and reconnaissance troop – and had no idea what chalk or helo they were on. I hoped they were all okay, but it was not the time to reflect on that, as we were going into battle. So, once again, I checked my kit – not because I needed to, but to get my mind focused on something else. Truth of the matter is that I have always hated helos. Once we had to do an emergency landing during training and something went wrong, injuring some of the lads. Nothing serious, but the feeling of utter helplessness when the machine lost control, even briefly, was enough to put me off them for life. Not only that; when up in the air, you're at your most vulnerable – something we all knew well. As did the people with some heavy hardware below us.

So we had lost one craft before even landing in Iraq. Sadly, we later learnt that all twelve people aboard that helo died in the crash. They were the first Allied casualties of the war.

The helo's green lights flicked and the loadmaster held up two fingers. Two minutes away. Clutching my weapon, I checked the safety. It was on. I pointed it to the floor, and slid it to fire.

As we were about to land, bursts of tracer fire lit up the refinery in a frenetic pyrotechnic display. It seemed as though the whole of Al-Faw was expecting us. We knew that the peninsula was hotly disputed territory with Iran and consequently heavily guarded, but this was pretty unnerving. Lead zinged everywhere. The loadmaster opened the back, and we ran out and took cover. Once the last Marine was off, the helo thundered back into the sky. I'm sure the pilots were spooked by the fiery welcome.

Navy Seals began breaching the refinery entrance, while 40 Commando spread out, ready to storm the place. Soon, engagements were taking place everywhere, as we began the mammoth task of clearing the pipelines and pump stations. We were significantly outnumbered, but what we lacked in feet on the ground was more than made up for in determination, weapon proficiency and training. There were concerns that even if we took the oil pipeline, we wouldn't have enough people to defend it, but we went ahead anyway. That's what Marines do.

We cut about from place to place, gradually securing section after section of the complex as we went along. There were numerous encounters; some nests of resistance were more tenacious than others, but we didn't lose a single firefight. Royal Marines are expert war-fighters – but I don't need to say that, as the rest of the world does. All I knew was that whenever shot at, I had brothers next to me, returning fire with interest. As I did for them. It's a bond few civilians can imagine.

When daylight broke, we had control of the refinery and – equally importantly – had thwarted any attempts at sabotage. There would be no ecological catastrophe on the Al-Faw peninsula. That alone was a massive triumph.

Outside the gates, some Iraqi women showed us pictures of men who had been defending the pipeline, asking in Arabic and broken English if we knew where they were. I recognised one in a photo being held up by, I assume, his wife. Sadly, he was lying just around the next corner, his AK-47 unloaded and kicked off to one side. He had caught quite a few rounds in our initial assault and bled out.

The next objective was securing Saddam Hussein's Ba'ath Party headquarters on the way to Basra, the country's second largest city and main port. All objectives had James Bond codenames for locations and attack points, such as Goldeneye, Moonraker and, my personal favourite, Pussy Galore, showing that someone in HQ at least had a sense of humour. So, with barely a pause, we hit the road for the next 007 'destination'.

By then most of the locals had fled and we 'appropriated' a complex en route to Basra to be our home for the night. It consisted of a cluster of buildings and 40 Commando was divided between the various compounds. We were all on edge, as our intel was that hardcore Baathists and fedayeen, fanatical Saddam supporters, were mounting up in tanks and heading south to meet us head-on. So far the fight had been relatively easy, with only foot soldiers to contend with, but it looked like things were about to change, with some heavy metal coming our way.

As a result, the anti-tank lads went onto the roof of the compound to set up the latest in armour-busting missiles and targeting systems. Operating as heavily equipped but highly mobile units, they had a solid reputation in the Marines for working hard and playing harder. Their kit was equally impressive. And expensive: one missile alone cost the price of a UK house.

Charlie Company joined the anti-tank lads on the roof that night to share sentry duty with a rota system of two awake while the rest slept. The weather suddenly turned and a mini-storm swept through our position. The hard rain and gusting winds only lasted ten minutes, but it had been unexpected and we were soaked to the bone.

Then, as the wind died down, we heard it: a low distant rumble – the unmistakable sound of a tank. The lad standing guard next to me was immediately on the radio net to the rest of the anti-tank team.

'Stand to, stand to, all call signs!' I could hear the guys below scrambling to get onto the rooftop.

He then slid over to the Javelin missile and I heard whirring sounds as he flicked on its thermal imaging.

'H, pop open those LAWs. Get 'em ready for the lads.'

The LAW 94 resembles two drainpipes, one smaller than the other, with a missile inside. To operate, you simply slide it open until it clicks into place, wedge it against your shoulder, locate the target and fire. I popped one open and threw it onto my shoulder. The sights are pretty basic, but every Marine is proficient in the use of these portable anti-tank rocket launchers.

The rumbling grew louder and most of the lads were on the compound roof, searching the horizon for a T72 in the hope of destroying it with a missile costing the same price as a bungalow.

'Soon as it's in view, let go on my command, lads.'

The tension was unbearable. You could have cut the atmosphere with a knife.

Then, suddenly, a shout. 'Do not fire!'

Then another. 'Stand down! Stand down! It's British armour!'

No one had thought to tell us that British tanks had been ferried ashore and were moving directly to our front across the battlefield. We had been nanoseconds away from taking out one of our own, which would have been a really nasty moment in Op Telic's history.

Luckily the anti-tank squads have reflexes like striking cobras and the lad with the Javelin knew right away what he was looking at. Even so, that tank literally dodged a bullet – or, more correctly, an expensive missile.

After the excitement of the tank episode, it was a relief to be on the move again. We secured three tall buildings on the outskirts of Basra as our next accommodation. They seemed to have belonged previously to the Iraqi government, as there were plenty of desks and chairs lying about. The lads claimed a room on the ground floor, facing the road, and we sandbagged the windows. As night fell, we took sentry duties on the roof; I got the 2 a.m. shift. This was good, as it was currently 9 p.m., so I could sneak in a few hours' sleep. I was dog-tired, as the last few days of fighting had taken their toll. The second I put my head down, I was out for the count.

Not much wakes me up quickly. But an RPG smashing into the front of a building I'm in certainly does. As I jumped up, my first thought was that I had gone blind in the blast, as I couldn't see a thing, but then I realised why – the room was filled with choking dust.

'Contact!'

We all rushed up onto the roof, each one of us covering arcs of fire across a field to the north. However, whoever had taken the RPG shot appeared to have done a runner. It's known in the Corps as a 'shoot and scoot', and it certainly gets the old adrenaline going, if nothing else.

Next, we secured the Ba'ath Party headquarters and orders came to continue our push north towards Basra Palace. Large Republican Guard mechanised units, equipped with T72 tanks, still lay ahead. Personally, I didn't fancy a one-on-one with a T72 – it would not be a good day out.

However, we had been promised air cover, and the headsheds were as good as their word. Forty minutes before we moved up on the position, attack aircraft unleashed a barrage of tank-busting shells on the Iraqi armour unit waiting for us. Even though we were some distance behind, we could feel the thuds shaking the ground, testimony to just how deadly those fighter planes were.

With the bombing over, we continued advancing. A couple of miles later, we came upon a scene of utter devastation. The Russian tanks of the Republican Guard had been completely obliterated. On one shattered turret lay the crumpled body of an Iraqi commander, his tank and the palm trees around him burning furiously. I felt as if I was on the set of some old Vietnam War movie – *Full Metal Jacket* or *Apocalypse Now*. To this day, it is one of the most surreal scenes I have witnessed in combat, and it told the story of the battle for Basra more vividly than any words.

Then came another command: 'Lads, stay away from the tanks; they're hit with uranium rounds.'

Depleted uranium armour-piercing shells are radioactive and could cause all sorts of health problems. I don't often do as I am told, but in this instance I backed off to my vehicle as fast as lightning. With me was the unit's doctor, whom we called 'Morph', short for morphine – how he got that name is a story in itself: during a previous exercise on Dartmoor, some lads had rolled an open-top vehicle and were in bad shape. Morph was first on the scene and, as some of the injured had broken bones, he pulled out his morphine pen to administer pain relief. Unfortunately, in the darkness he didn't notice that the pen was the wrong way around and the auto-injector went straight into his hand. By the time we got there, he was off his tits, sitting between injured Marines and laughing while trying to eat the rain. I don't think he will ever live that down.

The frontline of the Republican Guard was destroyed, but we were still pinned down by some fierce resistance, so command called in artillery fire to soften up the targets. Artillery back-up has to be a hundred per cent accurate, for obvious reasons, and there is a strict procedure for it. Somehow, in this case, it got misinterpreted and before we knew it, 40 Commando had airburst artillery rounds landing on our own position.

The call for help was instant: 'Casualty Charlie Company!'

A grid reference was given and, as the doctor and medics were already in my vehicle, I sped off.

One of the guys did the navigating and after a short drive we came up to a group of blokes lying on the floor. At first glance I could see three casualties: a lad holding his foot; another covered in blood with a FFD (First Field Dressing) compression bandage stuck to his cheek; and a third, out cold and bleeding heavily. We rushed straight to the third guy. He had severe shrapnel injuries and was bleeding internally with a collapsed lung. A rasping sound erupted with every breath. He needed a drain to clear the fluids congealing inside his chest cavity, or else he would suffocate.

The doc and the medic got him onto a stretcher and began working immediately.

'H, grab this,' said Morph, placing a clamp on the guy. I held the clamp with my left hand, keeping my right hand free.

'Right, now hold this tube.'

Metal clamp in one hand and the drainage tube in the other, I watched the Marine's face for signs of life. He was deathly pale from blood loss.

'Okay, here goes!'

A ton of blood and gunk came shooting out of the tube, unfortunately missing the container and splattering instead all over my desert boots. They went from the colour of sand to red in an instant.

'Great.'

'Lads, over here,' came another urgent shout.

A group of four more guys stumbled up to us, holding a Marine who had a piece of frag hanging down the back of his arm. His triceps muscle was peeled neatly back, looking like a kebab skewer.

We needed medical evacuation immediately and a helo was coming in from HMS *Ocean*'s medical bay. While we were tending to three critical cases, the lad making the most noise was the guy with the injured foot. I'm sure it was painful, but he was the least of our worries. You can live without a big toe.

I scanned the skies. Soon, we got the call we were waiting for.

'Lads, helo in two mikes,' meaning a helicopter would be with us in two minutes.

The doctor and medic frantically prepped the casualties for pick-up, wrapping up the worst hit Marine for protection against the hot downwash of dirt and dust from the incoming helo.

The Sea King came in low and slow. Our HLS (helo landing site) was clearly marked, but for some reason the pilot missed it, landing about 400 yards away.

'Someone get that pilot onto the HLS,' shouted the doctor.

I sprinted over as fast as possible, lungs on fire in the scorching heat. I was also in full combat kit and the helo rotors were thumping hot downdraughts directly onto me as I reached the pilot's side window, shouting, 'Follow me!'

He leant forward to say something, but I was gone. There was no time to waste, and he hovered in the air just above, as I pumped my legs as fast as I could to the HLS. I'm sure he was pissed off being given orders by a lowly Marine, but fair play, he did move fast. All casualties survived that friendly fire incident – an extremely lucky escape.

The day had been a busy one; sitting against the car, I looked down at my boots. The blood and gunk from the Marine's collapsed lung were more brown than red, but were kicking up a bit of a stink. I felt a bit sick and couldn't eat a thing, despite having had no food throughout the day. Although I had been running in the heat, I knew it definitely was not sunstroke. I drank more water, but still felt drained and nauseous. Many years later someone told me that reaction was in fact delayed shock. I never even knew there was such a thing.

After that, the Iraqi Army was either making a run for it, ditching their uniforms in the process, or surrendering. By the time the Yanks had secured Baghdad, Basra was pretty much wrapped up, including our target of Basra Palace. A disused factory, complete with abandoned Scud transport trucks, was 40 Commando's new home. The multi-wheel missile-launchers weren't the only things that had been dumped; there were piles of AK-47s, live ammo and grenades scattered everywhere. It was a munitions nightmare and our first task was to clear it all. There was no time to hang about, as people, particularly kids, would die playing with this lethal hardware. So we fired off all the ammo and simply dropped the grenades into the sewers – job done.

However, although we had won a military victory, there was zero civilian authority being enforced in the city. There was no police force to keep law and order, and we were commandos, not cops. As a result, the locals went on a looting spree that had to be seen to be believed. If something was not nailed down, it was stolen. I watched, amazed, as vehicles, TV sets, computers, furniture, boxes of tiles, building materials and even full kitchen units were brazenly carted off by locals.

We shrugged. Who were we to stop them? After years of brutal rule and persecution, Iraqis were making the most of their now lawless country. I guess it was their idea of freedom.

We all knew this was not going to end well. The power vacuum left behind by Saddam's ruthless regime was to lead to an equally sinister chapter in the country's history: the rise of an even more brutal insurgency.

9

Meeting Nicky

The first phase of Operation Telic was over and we flew back to Brize Norton in the dead of night to avoid the press. There were no welcome-home parties or flag-waving parades. The Royal Marines didn't go in much for that sort of thing anyway. Greeting us were just a few wives, back at Norton Manor camp, as the bus pulled into Taunton.

I said goodbye to most of the lads, who were leaving to go home to family and friends, as I would be staying behind at base. Lincoln didn't have anything going for me, and probably my biggest love was my pride and joy – the Celica. I had covered the car up back in December, when we left for the Gulf, so I connected the battery and the reliable Toyota started up first time. Driving decent cars at speed was one of the things I missed most when away, as military vehicles tend to be big, cumbersome and uncomfortable. It was superb to be behind the wheel of the Celica again and I thrashed her around the back roads of Taunton to do some quick shopping.

It was time for something I had been dreaming of for months while sweaty, sticky, unwashed, unshaven, and covered in desert dirt and grit – a hot bath and a cold glass of white wine. I lay in the water, barely moving, until it turned lukewarm and every part of me was wrinkly. I was thinking of meeting Jane that night, but was too shattered to be good company so decided to catch up on sleep instead. I was too tired even to dream.

Daylight woke me early, as I hadn't even bothered to close the curtains. Feeling refreshed and revitalised, I grabbed my phone. I hadn't taken it with me on Op Telic and, as I switched it on, it started beeping with a long list of messages, including some from Jane. I clicked on the first one.

'H, I hope you're okay and safe. I just wanted to let you know that I got a job in Southampton and have moved over there. If you want to come over once you're back, that would be great.'

That was the end of that. I wasn't in the mood to drive from Somerset to Hampshire, so instead dug out a pair of jeans, a hideous shirt and drove into Taunton. Nothing had changed, apart from the fact that it was not quite midday and I was already drinking. A few other lads were in the bar as well, as one thing about the Marine Corps is that you're never alone. You can go to any corner of the globe and will always find a bootneck in a bar. After a good session under my belt, I was happy, drunk but still jet-lagged, so I was back in bed before *EastEnders* had started on the telly.

I stayed in Taunton for the rest of the summer and, like most holidays, that time came and went all too fast. Then it was back to work.

However, Marine life after the first phase of Operation Telic started to get a little dull. Doing exercises in the UK, replicating what we had already done in the heat of battle in Afghanistan and Iraq, just didn't do it for me. Once you have experienced frontline action for real, everything else feels like a massive waste of time. I also had a major bust-up with the new regimental sergeant major who had replaced our previous one after Op Telic 1. We got off on the wrong foot right away, as one of his first chores was to deal with a bar brawl I had been involved in, and he hadn't handled it well. We had upset some Taunton lads over a girl, and a bit of a tear-up ensued. To be fair, we were all very drunk, but I got the brunt of the bollocking once back at camp. The RSM is basically the enforcer in a unit and responsible for all discipline, so falling out with one is not unusual. But it still didn't augur well for me.

The dynamics were changing fast, not only for me at the base, but in the Middle East as well. Iraq was kicking off big time once again, with the most shocking incident being the murder of six Royal Military Police in the village of Majar al-Kabir, 120 miles north of Basra. They had been cornered at a police station by a mob of up to 600 people, and beaten and shot to death. It was the biggest single loss of life since the war had begun and, like everyone else in 40 Commando who had been following the news, I knew we would be back there soon. Sure enough, when Op Telic Phase 4 was announced, it had our names all over it. For me it was to mark a new chapter in my life.

Unfortunately, the build-up to Telic 4 was shockingly bad and the operation itself was worse. We had been promised loads of Gucci kit, such as helo roadblocks for special forces to grab high-profile targets off the ground, but that never came off. We even had problems with ammo supplies and operations getting binned – and of course the RSM seemed to have it in for me.

The camps were also run by what we call 'camp commandos': people who rush around telling everyone what they can and cannot do, but never actually leave the safety of the base themselves or fire a shot in anger on a battlefield. These guys were only in Iraq because of what our unit had previously achieved. During the actual invasion, the lads had operated in hugely challenging conditions, usually outnumbered, yet still performed superbly. This time around, jobsworths were pulling me over for speeding on an airfield in what was supposed to be a war zone.

Within two days of arriving in Iraq, I had handed in my notice to leave the Corps that I loved so much. It was the way the op was handled that was the final straw for me.

I spent the rest of my time on that tour focusing on lining up a new career in close protection (CP), starting with an intensive bodyguard training course. My seniors made it as difficult as possible to get time off, perhaps understandably so, but luckily I had a bit of top cover, as my company sergeant major was also about to leave the Marines and do the same course. I was his guinea pig, so to speak, as he would get all the information and answers from me before he started his sessions.

Leaving the Marines was a huge, heart-ripping decision. This was, after all, my family – closer-knit than my blood family. The wrench is almost indescribable. It was all I had lived for since Corporal Jenkins had spoken to me almost eight years before at the careers exhibition at school. It was in my blood: something I was prepared to die for. I had grown up in the Marines and had a network of lads around me like no other.

I was about to leave that unique, absolutely magnificent brotherhood and move into CP – something that would have been unthinkable to me just a few months previously. But with Iraq once more in flames, private contracting was becoming massive, as governments were finding it more convenient to contract out security in war zones. It looks better in the media to have a contractor killed rather than a soldier paid for by taxpayers. Consequently, former rapid-reaction soldiers like me, particularly from the elite fighting units, were in big demand and paid good money. The going rate was seven grand a month in 2004, more than triple a Marine private's salary. It was not without risk, but I was single and carefree, so I rolled the dice and decided that was the life for me.

My life also changed in another way, and one I was not prepared for. It happened when a friend from Lincoln invited me up north for a weekend of drinking. I had not been back home for a long time, as I hated bumping into people from my schooldays. I tried to be friendly, but it always felt like I was from another planet, as I was far too wrapped up in military life.

Anyway, I reluctantly accepted that time and met up with my mate at a bar called Revolution, which served every type of vodka you could think of. Our favourite was cherry cola for adults and White Russians, both absolutely lethal combinations. We then started playing Russian roulette with clear vodka shots: ten are lined up, one of which would be infused with chili. It was a great game if you won, but a trip to the bathroom if you didn't. Luckily, I won that one.

Through a vodka haze, I saw a girl I recognised from school. Or, to be more correct, I saw who she was with

– a tall leggy blonde with curly hair down her back. She was beautiful. They were in a corner under the stairs, so I made my way over and made small talk with my school acquaintance. My real interest, of course, was her leggy friend.

'You guys want a drink?' I asked.

They did. I pulled out some cash and asked my school friend to get the drinks, pretending that I needed to go to the bathroom. As soon as she left for the bar, I introduced myself to the gorgeous blonde.

'Hi. I'm H.'

She held her hand out.

'Nicky.'

We chatted, and she told me that she worked for the council and lived in Lincoln. She had a really nice way about her, gently spoken, and an amazing smile.

'My friend said you are in the RAF,' she said, flashing that drop-dead gorgeous smile. 'But I said no way. You definitely don't look RAF.'

'Beautiful and intelligent,' I thought.

'Good observation,' I said. 'Royal Marines. Well, not for much longer. I'm leaving.'

We spent the rest of that night drinking and perhaps I overdid the booze, but somehow I got Nicky's number. Maybe she felt sorry for me, but one thing I did know was that if I got a chance to take her out, I would not mess it up.

The Marines gave me time off for 'resettlement', which was basically leave of absence to retrain in something for a life outside the Corps. For me it involved a close-protection course in Hereford given by four ex-SAS guys, highly respected in special forces circles. They knew their stuff and the excellent programme they had put together reflected that. For a month we lived and worked out of a giant manor house, honing skills on surveillance, counter surveillance, advanced driving and close-protection drills, as well as medical and trauma care. For me it turned out to be one of the best courses I've ever done and the skills I mastered on it have probably saved my life more often than I can remember. This was a pass or a fail course and I put everything into it. I had no back-up plan, so it was the life of a CP specialist or nothing. I passed with decent marks and got interviews with several companies, but as I had already picked the one I wanted, the process was an easy one. I chose an outfit that had won a massive reconstruction contract throughout Iraq and I already knew a few lads working there. The company signed me up, and the next phase of my life began.

But before heading to Iraq, my plan was to go to my grandma's house in Lincoln and take Nicky out on a date. She knew I was due back home, as we had been regularly messaging each other since a week or so after the night we met.

The date was perfect, just as I had imagined it. We went to the cinema and then out for drinks, sitting for hours on a sofa in a bar by Brayford Wharf and enjoying each other's company. She was great to be with, lighting up a room just by walking into it. We are opposites in a lot of ways, but have the same sense of humour, and she loved to mercilessly take the piss out of me.

We fell in love. However, I had not factored that into my new adventure – in fact, I had never even considered it. And the last thing I wanted to do was burden Nicky with the whole nasty Iraq situation. The barbaric savagery of the insurgencies waged by al Qaeda in Iraq (AQI) and the Mahdi Army was constantly on the news, and Nicky struck me as being a bit of a worrier, so it didn't seem right to make her even more anxious about me.

So I lied, telling her I was off to Kuwait to look after a rich family.

My tickets came through for Baghdad, from where I would be sent to another still-to-be-decided location to join my SET (Security Escort Team). It was time to go.

However, leaving Nicky was far harder than I had imagined. My last few weeks with her had been idyllic, and she totally accepted me for who I was. There was no need for me to showboat, as she was chilled out and not in the slightest bit interested in material things. She was as wholesome as they come. I knew I would miss her deeply. I knew she was the one for me.

She dropped me off at Newark Northgate Station – always a cold bleak place but made even more so as we hugged goodbye. It was a ninety-minute journey to London, and from Heathrow I would fly to Jordan and then Baghdad, returning to Iraq for the third time in two years. But this time I was a civilian contractor and making the kind of money most twenty-four-year-olds could only dream about.

The security company had large headquarters in the Iraqi capital which became my home for the next ten days. Among my new teammates was a big, dark-haired ex-Royal Marine called Merrick. Being former bootnecks, we gravitated towards one another. He had been a bodyguard to David and Victoria Beckham, and had loved working for them, but the lure of contracting and the ridiculous salary had gotten the better of him. Indeed, as it had with most of us.

We learnt how the company operated throughout the volatile country, and what was expected of us, and familiarised ourselves with new weapons and equipment. The company was very well run, and even had a legal team to brief us on use of force and other sensitive issues.

Upon completion, we were asked where we wanted to operate. I chose Fallujah, as I had mates there and it was where most of the action was at the time. The Yanks had gone into the city to reclaim it from hardcore insurgents, mainly AQI and fanatical Baathists, and it was a brutal street fight. I wanted a piece of that.

Merrick opted to stay in Baghdad. We said goodbye as my transport to Fallujah arrived.

'Be careful, H, it's lethal up there.'

'And you, bro,' I replied. 'It's not exactly paradise here.'

We shook hands. It was the last time I saw him. A few months later his vehicle was hit by an EFP (explosively formed penetrator), a type of molten lava projectile that went through armour, including that of tanks. Merrick was killed instantly, as was Chris Butcher, a former Army Commando.

It was a sombre introduction to my new world.

10

The Triangle of Death

You can't really talk about the Iraq War without mentioning Fallujah and Ramadi. These were hairy places even for the military, with their massive logistical support, so imagine what it was like for us, in eight-man private security escort teams with three cars.

Fallujah was probably the most infamous hotspot, located in what was referred to as the 'triangle of death'. The city was a simmering volcano of resentment, violence and outright hatred, with Sunni insurgents in control. For Westerners, setting foot there meant certain death in the most barbaric manner, as four Blackwater security contractors, who mistakenly turned into Fallujah on 31 March 2004, discovered. They had driven off the main supply road called Route Mobile, and once jammed in by traffic, the insurgents swarmed all over them, first disabling the vehicles with hand grenades. The four contractors were beaten to a pulp before being killed and set on fire. Their mutilated bodies were then dragged through the street by the jubilant mob – a gruesome re-enactment of the 1993 'Black Hawk down' incident in Somalia – and eventually strung up on the bridge over the Euphrates River for the world to see. It was a horrific warning of what was in store for foreigners.

We watched the gory video on the Internet, as the jihadists had taken great pride in filming every grisly second of the murders. The brutality was staggering and, unsurprisingly, Americans were outraged, both in Iraq and in the States.

In April 2004 the First Battle of Fallujah began, with the Americans vowing to destroy the Sunni insurgents' stranglehold and also to get the killers of the Blackwater contractors – no mean feat in a city of 300,000 people. It was a huge op, with the US Marines leading the charge. It ended in a stalemate, as it quickly became apparent that the Americans would have to rethink their simplistic strategy of booting the militants out.

The Second Battle of Fallujah started six months later. This time the Americans studied the city layout and the insurgents' tactics – how it would have to be taken street by street and correcting the mistakes from the first attempt. Nothing helps things to evolve faster than warfare.

Within three months they had secured the city. But that still didn't make Fallujah safe. The triangle of death was not called that for nothing.

Camp Fallujah was my new home, and it housed Marines and Navy SEALs who had fought in both battles. I even bumped into a SEAL who, like me, had been on Operation Telic, breaching the pipelines at Al-Faw.

We soon became friends with the US Marines and, being a former Royal Marine, I could easily relate to them. We looked after one another and if they needed something that they could not get from their camp, we would happily bring it in from another location. In return, we got on-the-spot intelligence and kit – not a bad deal, if you ask me.

We were involved in a couple of IED (improvised explosive device) incidents in Fallujah, as well as a few 'shoot and scoot' ambushes, when the attackers fled the moment we returned fire, but nothing too serious, considering the war footing the city was on. But as Ramadi, another hotspot thirty-three miles west of Fallujah, was now also flaring up, I was moved to a Security Escort Team called there. Ramadi is the capital of Al Anbar province, the centre of the triangle of death, and – like Fallujah – it sprawls along the Euphrates River. I knew this was going to be an 'interesting' posting.

The company had two SETs in Fallujah, but only one in Ramadi, so we would be pretty busy. My new team, SET 20, comprised eight guys. Daryl Fern, a Scotsman, was in charge, with a cockney lad called Shaun Barber as his second in command. Both were great leaders, and I was lucky to be in a very close-knit and efficient team, particularly as we were based in the most dangerous area in the world.

Daryl soon discovered I had been a specialist driver in the Royal Marines and tasked me to drive him in vehicle one, the command vehicle. That suited me fine. Driving in the death triangle was fast, furious and formidable, not least because the armoured vehicles were top-heavy and difficult to control at speed. The last thing I wanted was to die at the hands of someone else's poor driving, especially when the end result might well be insurgents mutilating our bodies and tossing them off a bridge.

Our new home was Camp Blue Diamond, on the outskirts of the city, built on a sliver of land that also housed a contingent of US Marines. It was apparently one of Saddam Hussein's son's palaces, but we would not have guessed that, as our new living quarters were anything but palatial. The previous occupants had been a well-known American private security company, which had been unceremoniously kicked out of Iraq after crossing a few lines. They had left the place in a total mess. For some reason they had taken to scribbling provocative graffiti everywhere on the giant concrete T-walls, the most cheerful being, 'Only the dead have seen the end of war', accompanied by grim-reaper drawings.

'H, you're in room eight, bro, opposite me.' Daryl chucked a key in my direction. A commander and his driver were always close by, in case of emergency moves.

The previous contractor in my room had clearly been a frustrated artist, as sketched across the entire wall was a scene of the infamous Blackwater killings in Fallujah. The guy had even painted the bodies dangling off the bridge, complete with slogans and vows of revenge. I figured he must have known the murdered men. But even so, the last thing I wanted was to live in a room with grisly graphics of murdered people, so I had to sort the mess out fast. Finding paint in a war zone military camp is impossible, but I did find a load of camouflage ponchos, which I nailed over the dubious work of art.

However, with all the maps we had of the city and the torturously roundabout routes we had to take to minimise danger, my 'poncho' wall soon became something far more useful: a detailed chart room, showing plans, roads and grids of the entire city. This became incredibly useful in plotting our journeys and avoiding ambushes.

It wasn't long before we were introduced to our client, an American team of engineers who would oversee reconstruction projects throughout the area. Some sites were easy to get to, but others were in the middle of deadly no-go zones, requiring intricate planning or simply not going at all. Our first ports of call were the many bombed-out bridges, as re-establishing supply chains was the key priority for the shattered country. The main supply road, Route Mobile, was officially the most dangerous motorway in Iraq, as it passed through most of the major trouble spots. It wasn't protected and driving along it with civilian clients was a calculated gamble with people's lives, but we had no other option.

However, one of our riskiest runs did not even involve clients. We operated on a monthly rota system of eight guys on duty and four guys on leave – something vital in such intensely stressful working conditions. This meant that we had to take four guys out to the airport, before returning with the four who had just arrived back. This was an eighty-two-mile hell run, starting at Ramadi then skirting past Fallujah and on to Baghdad. It also had to be done in daylight, as no one wanted to be out anywhere in Iraq after dark, let alone on the IED-strewn Route Mobile.

Our first trip set the tempo for all future runs. It had to be an early start, which for me involved lots of coffee, and for Daryl, cigarettes and diet cokes. He lived on tobacco and low-calorie sodas, with a stash of cans hidden away in every spare compartment of vehicle one. Paul, a former Royal Marine and close friend, was our gunner in the back, manning a belt-fed Minimi machine gun that faced out through rear portals that popped open. The third vehicle – or gun truck, as we called it – was commanded by Shaun, while the client vehicle, driven by an Irishman nicknamed 'Goat', was the middle car. Luckily, on this trip we were not carrying any clients, just the team.

Between Ramadi and Fallujah is the Abu Ghraib prison, another infamous landmark, and on the side of the road was a jungle of overgrown marshland reeds. This was a prime ambush spot and the insurgents loved it. They regularly hid out there, just waiting to prey on some unsuspecting foreigners.

That morning, we were the prey. A huge bang thudded across the road and vehicle two came up on the radio.

'Contact, contact! Vehicle two! No casualties and car is good!' shouted Goat.

'Drive, drive, drive!' shouted Daryl.

I saw a few AK-47s pop over the top of the reeds, but it seemed as if the gunmen were holding the rifles above their heads and firing blind. There was no way they would hit us like that, so their plan was to disable us with the roadside device and then draw us into the killing zone. Cool as anything, Paul laid down fire as we pulled away. A good guy to have on your side.

'You hit anyone?' Daryl asked Paul. He would have to write a report, and casualties had to be logged and noted.

'No, all good.'

Whew! Even when you want to take out the bad guys, the paperwork isn't always worth it.

It was also pretty obvious why the insurgents loved that position. It provided superb cover, as you could hide a platoon in there. The only reason I'd noticed the AKs sticking out was because the bullets were splintering the reeds. I never saw anyone actually holding a rifle.

We made it to the airport in one piece. However, even though vehicle two was still running, it was smashed to pieces and would be consigned to the scrap-metal graveyard at Camp Victory, the coalition's main site at Baghdad airport. That we got it there was testament to Goat's driving skills. A real character, Goat was usually found by our camp's makeshift swimming pool, wearing next to nothing and suntanning.

This was my first hair-raising airport run, and it opened my eyes a hell of a lot wider. From then on, we established a strict routine, leaving Ramadi before sunset and spending the night in the big Camp Victory transit tents. We would then get the off-duty lads on their flight out in the morning, and hang around until the guys returning landed in the late afternoon.

The return trip became known as the 'run for the sun' and was extremely hazardous. However, for the guys coming back to start their next tour, the adrenaline actually began pumping the moment the plane came in to land at Baghdad airport. To avoid being shot out of the sky, the pilot would 'corkscrew' with gut-wrenching turns, resulting in some epic cases of airsickness, not to mention abject terror for anyone with a fear of flying.

Still pumped, our guys would rush off the plane, kit up in body armour and weapons, and jump in the vehicles to chase the rapidly setting sun up to Fallujah. There, we would drop off the first batch and then head for Ramadi at breakneck speed, in darkness, which is not an experience to be savoured. The basic rule was: the faster you drove, the more daylight you had. Night was the enemy and there were some pretty bad people with itchy trigger fingers waiting out there in the dark.

Despite this, certain companies continued to run night convoys and paid the price. Most mornings we would pass at least one truck or an American Hummer in flames on Route Mobile.

Camp Blue Diamond was a great place for us, as there was no management ordering us around and a big DFAC (dining facility), where we could eat almost anything and as much as we liked. I'm not at my best first thing in the morning, so always had an early coffee and omelette for breakfast, preferring to take my time with any mission prep or route planning. As the vehicle-one driver, getting to and from sites was my responsibility, although Daryl would navigate. But with all the maps on my wall, I usually knew exactly where I was going.

One morning we were on a mission to a Primary Health Clinic (PHC) which we had not visited before. A new engineer had arrived in the country and was obviously keen to impress his bosses. He was a skinny guy who told us how he used to be fat, but had become vegan and turned his life around. He was a likeable fellow, but somewhat naive; he also wanted to change the lives of the Iraqis and make a real difference to the country.

We explained to him the night before that this PHC was in an extremely hostile area and we would have to pass two no-go zones just to get there. That didn't deter him. 'We must go,' he said. As he was the client, we decided we would have to risk it.

The morning was clear as we headed north on Route Mobile. However, we had to cut across a 'soft road' about eighteen miles further up. This basically meant a sand track that locals used, as opposed to the tarmac road. A crazy thing about Iraq was that it had massive three-lane motorways, but every now and again the barrier in the middle was removed and it became six lanes of clogged two-way traffic for a good few miles. Someone once told me this also acted as an emergency runway, so Saddam could land his private plane anywhere in the country if he needed to. I'm not sure if that was true, but it sounds like the type of thing a paranoid dictator would do.

However, the soft roads, which many locals preferred – perhaps in case Saddam's plane landed on them – were not clearly marked. In fact, they weren't marked at all on any normal map, so finding them was difficult. Daryl counted us down every few hundred yards, as we approached the sand turn-off, and then shouted, 'Turn left.'

I swung the wheel and the Toyota lurched over the kerb onto the desert.

'One's on,' Daryl said over the radio.

'Two's on.'

'Three's on.'

Checking in his wing mirror to confirm we were all off the tarmac, Daryl nodded at me to up the speed. Even though we had not been working together long, we seldom needed to speak. We were on the same page with everything.

The desert road was ridged by sand berms and we hadn't got too far when I noticed something looking decidedly strange on the left-hand side.

I pointed it out to Daryl. 'Heads up, dude.'

He reached for his M4 with a times-four scope. Scanning though the telescopic sight, he suddenly exclaimed: 'Fuck me! It's an execution.'

I slid my M4 out and also took a look through the scope, using my forearms to lock the steering wheel. Face down on the berm wall were five bodies with hands tied behind their backs. A large area of sand around them was soaked in blood. Judging by their uniforms, they were policemen working for the Americans, trying to establish a new Iraqi government. Pinned onto their shirts were notes written in Arabic, denouncing them as traitors. They had been lined up and shot in the back of their heads.

'Push forward and let's get eyes on,' said Daryl.

I drove slowly, looking for booby traps. Insurgents regularly rigged IEDs or grenades on bodies.

'No wires, lads. No devices,' I called out

'Earth around them hasn't been disturbed,' said Daryl. Digging was always a tell-tale sign of a hidden bomb.

Daryl and I had checked all of this out from vehicle one, while the client was kept out of the way in vehicle two and the gunner watched our backs in vehicle three. We didn't want the client to see this.

'Okay, lads, push past, but stay in H's tracks,' said Daryl as we moved off.

They obeyed. Iraq wasn't the place for heroics and bravado. If you were told to follow someone's tracks on a road that could be rigged with IEDs, that's what you did. Everyone just wanted to stay alive, make some money and keep limbs attached.

We arrived at the site of what was to become the new PHC, our client's project. Daryl and I were on edge after the gruesome execution scene and were first out of the vehicles, M4s wedged in our shoulders and safety catches off, as we scouted the area. It did not take long.

'Clear, boys,' said Daryl. 'Now debussing the client.' He signalled for him to get out of vehicle two.

Shaun and Will from the gun truck were also out. We had all-round protection with belt-fed weapons at twelve and six o'clock positions. It was the first time I had seen the client that morning and he was staggering around, ghostly pale. If I didn't know better, I would have thought he was drunk.

'The fuck's up with him?' I asked Daryl.

The client was looking vacantly at us, as if in a trance.

Daryl followed my gaze. 'Jesus!'

He headed over to Marty, the client's personal protection officer. Marty was an ex-army and ex-prison officer – a short, stocky, hardcore old-school Scotsman and the 'dad' of our team. We all respected him and, as an older guy, perhaps he understood the client's unusual behavior. We just thought, 'What the fuck?'

I saw Daryl and Marty move out of earshot of the client. Daryl then ran up to where I was, crouching behind a pile of bricks I was using as cover.

'He's in shock, dude. Seeing the executed bodies has messed him up. I'm going to pull it, H. Go grab the car, mate, we're pulling this one here and now.'

I got the car ready, while Daryl phoned ops to let them know the situation. They agreed that we should get the distressed client back urgently, and Daryl gently told him that inspection time was over and we had to go. He never said a word as we ushered him like a child into the car. This wasn't just a different planet to him; it was a faraway galaxy. Thinking Iraq was chocolate boxes and roses, and he was going to make a difference, was not going to happen.

The journey back was a lot quieter than the one in. We knew we had to drive the client back past the gory execution scene, so the lads put him in the opposite side of the vehicle.

I never saw that client again. He flew out almost immediately and I suspect he had some form of counselling or therapy. Whatever – but for him it had been a stark lesson that in the real world things aren't always as simple as group hugs. After dropping him off at Baghdad airport, we shut him out of our minds, as we had more practical matters to consider. In my case, what to do with my leave soon coming up.

I was really looking forward to seeing Nicky again. She knew where I was based, as I had stupidly left the download of my e-ticket to Baghdad open on my emails. So bang went my story about a soft job guarding a rich family in Kuwait!

We Skyped each other on most days, but I never discussed the dangers of what I did for a living. The last thing I wanted was her finding out the truth of what it was really about – like IEDs, firefights, runs for the sun and a client going into hypnotic shock at the sight of blood-soaked bodies. I didn't want her to worry, particularly as street shoot-outs between US Marines and insurgents in Ramadi were happening most nights. We had ringside seats, as over the bridge from our compound was Camp Hurricane Point, where the city pretty much began.

The US Marine in charge of the night shift guarding our section of the compound was a sergeant called Smithy, a massive guy and as mad as a box of frogs. He was the 'grandfather' of Blue Diamond and knew everyone in the camp, high-fiving anyone in sight. If it got hectic up on the roofs, with too much incoming fire, we would go up to help. He would just shake his head and say, 'Bad guys, man. They're everywhere.'

Sergeant Smith was not wrong.

What is now called the Battle of Ramadi is said to have kicked off in June 2006. But I can tell you categorically that it began way before then. I knew Navy SEALs and other special force operatives going deep into the city at night and fighting almost continuous street battles. In fact, I ate breakfast with them most mornings, and our team supplied them with kit that was difficult to come by. These informal meetings were called the 'shark base' and they provided gold-dust information which few others were privy to.

By then our team could travel only to the outskirts of the city, as going any further was considered suicidal, so it cut down on the number of missions. Even the Marines were calling in JDAMs (Joint Direct Attack Munitions) virtually on top of their own positions. It was a self-perpetuating mechanism of war – roadside bombs by them, and night raids by the Allied forces. It kept us awake to begin with, but after a while we got so used to it that on one occasion there was a shoot-out close to our compound and the whole team slept through it. We only knew about it when Sergeant Smith told us at breakfast.

Smithy always referred to us as 'crazy Brits', usually followed by his weird mechanical-sounding laugh, but I like to think that he said that as a term of respect, as he was a true warrior himself.

Another real American hero sharing the camp at that time was a SEAL whom the locals called the 'Devil of Ramadi'. His name was Chris Kyle and he is probably the world's most famous sniper after the blockbuster movie of his life called *American Sniper* was made by Clint Eastwood. One of his most extraordinary kills was when a car bomb was accelerating down a street towards a Marine checkpoint. There was no escape and the only hope for the men at the checkpoint was Kyle, based on a rooftop some distance away. He fired a single shot at the speeding vehicle and hit the driver smack in the eye, killing him instantly – an almost impossible feat. In fact, I wouldn't have believed it if Smithy hadn't shown me the photograph.

Kyle was credited with more than 150 official sniper kills (unofficially, it goes as high as 360), but the number of lives that he saved by taking out suicide bombers probably far exceeds that. Even though he survived the war, he was killed at a shooting range back home, while assisting a veteran suffering from PTSD. A tragic loss, but one thing I do know is that the incredible stories of the 'Devil of Ramadi' are all true. I was there for some of the time.

Ramadi was eventually locked down in June 2006 and all roads leading into it declared out of bounds. Suddenly, this was big news, coming soon after the Battle of Fallujah, and reporters flocked to the latest headline hotspot. They were desperate for war stories, and I sometimes wondered if that was all they saw. The bigger picture was what a mess Ramadi was and how terribly the civilians were suffering. But I suppose that would only be worth a couple of minutes at the most on CNN.

Thankfully, as far as I was concerned, there were no Brits in Ramadi. Apart from us, that is. This meant no

UK media coverage. So at least I knew that Nicky wasn't seeing on TV what I was seeing in real life.

$\star$ $\star$ $\star$ $\star$

Despite the rapidly increasing amount of work in Ramadi, the team always had time for banter. It kept us on our toes and also was a welcome release valve for all the tension. The maestro of the banter was our crack Minimi gunner, Paul. Like me, Paul was a former 40 Commando Royal Marine and we had been on Op Telic together, although I was attached to Charlie Company and he was in Alpha. For some reason, Alpha was known as the 'Saints' — The Saints! More like loveable rogues, a fine body of men that always looked up to us Charlie company boys! Not as gods, more like demigods! Paul would grab any opportunity to get us laughing, one such time bore the brunt of his wicked humour. I once foolishly told him about my fear of spiders, and then forgot I had done so. But Paul doesn't forget, and one afternoon I returned from lunch to find the rest of team sitting outside our accommodation, close to my room, joking around and watching me carefully. I joined in for five minutes or so, shooting the breeze, and then unlocking the door, went inside. As I walked towards my bed, the biggest camel spider I'd ever seen was lying on my pillow, staring at me. I fell over myself bolting out of the room at Olympic speed. The team outside practically wet themselves laughing. I didn't know that the spider was dead and had been strategically placed on the pillow, which made my rapid and undignified retreat even funnier as far as they were concerned.

The banter, like the missions, was non-stop, and a welcome relief from the constant threats we faced. The pranks were usually orchestrated by Paul, and the team was always up for a laugh. Paul's best shenanigans were saved for when I was paired with him on a mission. He spent most of the day chuckling to himself and asking weird, random questions, such as, 'Have you ever tried goat's milk?' and, 'Do you like goat's cheese?' I was a bit puzzled, but figured it was just Paul being Paul and I didn't think anything more of it.

But everything suddenly made sense when I opened my cabin door to be met by a huge, smelly and very much alive goat, tied to my bed and enjoying watching the movie *Windtalkers*. The goat turned its head to look at me, bleated and then went back to watching the movie. How the fuck? What the fuck? And *where* the fuck had he got this goat from? And *who* the fuck would clean up all the goat shit?

At that moment, a roar of laughter came from the lads outside my cabin, 'Hey, who do you have in there, H? Is it Vincent van Goat?'

'Oi, Billy Jean King! Come on out!'

'*Joseph and His Amazing Technicolor Dream Goat!*'

Paul shouted, 'Ask it how it goat there, honestly!'

You can picture me having to untie the goat and pull it out of my room while it carried on leaving its stinking piles of gifts for me!

'For goat's sake, H, stop creating such a ruckus!'

'Here we goat again.'

Yes, if you are wondering, I am still traumatised to the extent that every single time I see a goat, or food label mentioning a goat, I start laughing! No goats where hurt,

I should say, while spinning these dits – Marine slang for a story.

Apart from being a great operator, Paul was always upbeat and full of an energy that played a great role in keeping team spirits high during rough times. He is a living legend, a great friend and someone I can always turn to for help – and a bloody good laugh.

11

The Ramadi Run

It was time to head home for a month's leave again. Part of me wanted to stay and see how the Battle of Ramadi played out, and part of me longed to be with Nicky. I wanted the best of both worlds: the gorgeous girl and the thrill of a war zone.

Getting home always involved alcohol. Lots of it. We would land in Jordan with five hours to kill before the 2 a.m. London flight. After twelve weeks of fizzing adrenaline, we hit the bar hard, and continued drinking on the plane. By the time we landed at Heathrow, most of us were in a state. I would then catch the train to Paddington, have a beer in a bar, and head on to King's Cross Station, where I would grab a four-pack of beer for the last leg of the journey home to Lincoln.

I'm sure my wife could write the next part of this story a lot better than I can. On leave, drinking was what replaced the thrill of war. It's that simple, and it made me behave like an idiot. Booze gave me a god complex, and being both drunk and arrogant was a terrible mix. Nicky refers to this period as 'the time she nearly left me times ten'. I wish I could go back and change things.

By that point we were living together in a three-storey townhouse with a small garden that I had just bought. She is a saint. She would wait patiently for three months, worried sick whether I was safe, and then, when I finally got home, I would be blind drunk and acting like a complete fool.

Despite that, our relationship somehow continued to thrive, as she could cut through my crap. She also had a massive amount of patience and genuinely cared for me. I knew that and didn't want to throw it away. It had been a long time since I had been in love, and even in the chaos of my life at the time, it was a good feeling.

Nicky worked on our house and made it into a lovely home, although it was probably too big, as we only used the top and bottom floors. The middle floor was just a dumping ground for kit. But still, most of my time back home from Iraq was spent jetting off on holidays, as we were young and had money. However, Nicky was like a female Mr. T in *The A-Team*. She had to get drunk to fly, as it genuinely terrified her.

Downtime always went quickly, and in a blur I would be back at Heathrow Airport, bound for Jordan. When on leave, I checked in daily with Daryl to make sure the guys were okay, as our team was a family with an immensely tight bond. Travel to the Middle East was second nature by now: sleeping all the way to Jordan, catching the military flight to Iraq and gripping the seat belt tightly as the pilot corkscrewed into Baghdad. From there it would be jumping straight into vehicle one to run the gauntlet to Ramadi.

Daryl had a decent taste in music and his iPod was always plugged into the car speakers. He often joked that if we were hit and the vehicle went down, I was to salvage the iPod before saving anyone else. On these runs for the sun, he always played Phil Collins's 'In the Air Tonight' – nothing like a bit of dark humour as you're about to speed off down the world's most notorious road. Our motto, 'movement is life', became 'music is life'. If you didn't laugh, you would probably cry.

The Battle for Ramadi had grown fiercer in my absence. Every night I heard Sergeant Smith getting more and more animated, barking orders at his young Marines on the roof, manning belt-fed weapons. Insurgents often attacked the back gate, but it was more of a distraction technique to keep the Americans occupied while they planted IEDs elsewhere on the roadside. Even so, under Smithy, the Marines were learning from a maestro.

In the city itself, the battle had reached a tipping point, with US Marines dotted about at various outposts, trying to wrest back control of an area in which half a million people lived. It was a monumental undertaking, and as this was in June, it was being fought in temperatures nudging 50°C. That was bad enough, but as another contractor team had been hit with a VBIED (vehicle-borne improvised explosive device) that turned their vehicle into a flaming torch, we all had to wear Nomex fireproof suits. These were good bits of kit, but with body armour on top it was like being locked in a sauna on steroids. Every day was a hot and itchy sweat-fest.

Despite the heavy fighting, US Marines of Kilo Company had secured the Ramadi Government Centre, a vital building from both political and tactical standpoints. This was a location that one of our clients wanted to visit, so we had to get him in. Planning the course was easy, as there was only one main road in and out of the city, which

The car park in the compound with the embassy cars on the far left and abandoned suitcases in the foreground. This was the first attempt to get to the airport, using our own vehicles, which had to be cancelled after the Taliban surrounded us.

Luke Griffin (on horseback), who was killed in the Camp Anjuman attack in November 2018. Luke ran towards the explosion, engaging the attackers and slowing them down, saving countless lives as he did so.

Taliban entering the compound for the first time. The team observed them from the rooftop through night-vision goggles, counting down the seconds until they reached the meeting room.

The author (left) with Shaun Barber, just back in from a Ramadi government centre run, in Side Camp Blue Diamond.

A bad day at the office in Iraq, moving from Fallujah to Ramadi.

Daryl, Tracie and Ciaran on holiday. Daryl Fern was a natural leader and the SET 20 Ramadi team leader. He was killed by a roadside IED on 12 June 2008.

The Crossed Swords of Baghdad, Iraq, at sunset.

An early morning helicopter landing. Chaff, a missile counter-measure to confuse and distract, would be used as insurgents would often attack helicopters on their final approach to Ramadi.

The aftermath of the German Embassy bombing next door. The window blew in on me as I sat eating breakfast.

Paul Moynan, a Ramadi team member at the Crossed Swords in Baghdad during team training. It offered a wide open space that was great for vehicle anti-ambush drills.

An incoming sandstorm in Al Anbar province, Iraq. Sandstorms would impede missions as there would be no air cover or medical evacuations. If you were already out on the ground, sandstorms could cause some serious problems.

The HQ coffee machine acquired by my team. The Black Rifle Coffee Company coffee was an absolute game-changer – caffeinated and motivated, team morale was boosted by finally getting a decent hit of coffee.

The author and Daryl outside the operations room in Blue Diamond camp, Ramadi, in 2006. We had just finished a recce for a site we deemed too dangerous to visit with clients. Daryl has a much-needed cigarette in his hand!

The Airbus A400M in which the team flew out of Kabul on 21 August 2001, headed for Dubai, manned by 70 Squadron, Royal Air Force. It was practically empty, save for a few stragglers.

The author during the extraction from the embassy to the compound. It was a long night. The vehicle in the background had its back end ripped off during the speedy extractions.

An Afghan woman and her children on the way to the same plane we flew out on.
She sat opposite me.

Daryl (team leader in Ramadi) and Shaun (second in command in Ramadi), photographed in early 2006. They had a unique bond and led the Ramadi team through some challenging situations. They were an inspiration to all who worked with them. I would use everything I learned from them during the six days we were surrounded by the Taliban.

Staged and ready, the embassy B6 armoured cars used for the extraction. This was halfway through the night as the team grabbed a coffee as they waited for the go.

the Americans called Route Michigan. Timing, however, was key and the safest window of opportunity was during the morning call to prayer. Iraqis are deeply religious and the streets were the emptiest when the mosques were full. That's when we decided we would leave.

All Marine checkpoints were notified of our visit and intended route, and we had an American radio set to communicate with them. For some mad reason their call signs all began with 'Bastard'. I later discovered the unit's proud nickname was the 'Magnificent Bastards'. It was a good name.

At 3 a.m. our vehicles lined up at the back gate of the compound in inky darkness, waiting for the call to prayer from the muezzin as the sun broke. The cold metal of my M4 assault rifle wedged between the centre console and my right thigh offered some reassurance. My rifle was always made ready. In the car I had the safety on, and would flick it to single shot as I got out. We were most vulnerable when exiting a vehicle.

Waiting in an armoured car and about to rumble into the most lethal city anywhere on the planet played havoc with our nerves. Daryl had already smoked half a packet of cigarettes and I could sense this tough, battle-hardened veteran was as edgy as I was.

Then we heard it. '*Allahu Akbar!*' The call to prayer.

Daryl slammed the vehicle's doors shut as I keyed the ignition.

'Vehicle one's moving,' Daryl called over the radio.

As we cleared the gates, we got confirmation that vehicles two and three were behind us.

I floored the Toyota's accelerator. I had to thrash the life out of it to get up speed, as the standard factory engine was not designed for the weight of a fully armoured chassis. We crossed the bridge and hit the roundabout outside Hurricane Point, turning left onto Route Michigan. It was deserted while the faithful were praying, as we had anticipated, but that didn't mean it was safe. You were never safe in Ramadi. Not for one second.

There is a mind-boggling maze of side roads crossing Route Michigan. Our gunners continuously swept their Minimi machine guns from side to side. Insurgents often fired RPGs down those narrow alleyways, as they offered cover and excellent vantage points to attack convoys.

We had good comms with the Yanks and told them we were about to reach Checkpoint 295, a roundabout with a few Marines in a compound. This checkpoint was regularly attacked and the premises had been shot to pieces, but the 'Magnificent Bastards' held firm. Those dudes were warriors. They cleared us through and towards the Ramadi Government Centre gates. As I slowed down, I noticed that the car park in front of the building was flooded in knee-deep water. Or what I assumed to be water … until a Marine waded over and said, 'Welcome to Shitsville, guys. Park around the back.'

The stench in the desert heat was unmistakable. And unbearable. I realised why he had welcomed us to 'Shitsville'. He was knee-deep in raw sewage.

We drove around the building, setting up an overwatch position to provide covering fire if needed. We then escorted the client out of vehicle two and rushed him into the main building.

It was chaos inside. We had to navigate our way around unwashed, unshaven and exhausted Marines sleeping on camp cots in the narrow hallways. Even under the stairs.

They were clearly on a punishingly hard routine, constantly on standby, weapons never further than a grab away to defend a target relentlessly under attack. I could tell just by looking a few in the eyes that they had been seriously put through the wringer in this hellhole. They had an aura of war about them. These youngsters were true seasoned fighters.

Our client's meeting was in the Ops Centre, which sprawled across the entire second floor of the building. We went up and were greeted by a giant of a man who filled the doorway. He introduced himself as the CO and took our client to the spare room to talk. Daryl left two of our guys at the door and the Marines took the rest of us on a tour of this 'Alamo' in Ramadi's anarchic heart of darkness.

On the roof we saw the true picture of the city's destruction. It was quite a sight. Insurgents had cynically used civilian buildings surrounding the Government Centre as firing platforms, and most of these had become piles of rubble. In fact, the building opposite had taken a JDAM hit only the day before. After constant bombardment from this location, the Marines had had enough. They called in a strike to level it, hopefully with most of the rubble coming down on top of the insurgents. Unfortunately, the JDAM did not detonate, but it still destroyed the building, with bricks and concrete going up in a huge plume of dust.

We gave our client a maximum limit of forty-five minutes to wrap up his business, as the city residents would be done with their prayers by then. We needed to be off the streets before business resumed as normal – in other words, chaos. As the deadline ended, we rushed back to camp just as the call to prayer period ended.

Another successful mission completed, and in the debriefing afterwards, all agreed that the best time to do ops was when the muezzin cried, '*Allahu Akbar*'.

Our timing was fortuitous, as the Government Centre took another heavy hit soon after our mission, when about a hundred insurgents attempted to overrun it. Our American friends then realised that this running battle for real estate was not working. Many Marines had been injured, and some killed, in the fighting. Manpower was becoming an issue, not to mention the appalling living conditions in places like Shitsville and the constant supply issues. They decided on a new strategy, which, simply stated, was to level every building surrounding the Government Centre. It was effective, as insurgents no longer had any cover. They were shot as they approached.

Usually on Fridays we were off mission, so on most Thursday evenings the team would wind down from the week's events, sitting outside our accommodation in a circle, re-acclimatising and talking a lot of rubbish. The eight of us were the only Brits at Camp Blue Diamond, so often visitors came over to say hello or hang out. A lot of Americans have never travelled outside their own country and were curious to meet us, fascinated by our accents, and wanting to know more about London and life in the UK. There were many unusual questions, such as at what time we took afternoon tea or if we knew the Queen – like we lived next door to her. I imagine we disappointed them as, far from being mates with Her Majesty, we were hairy-arsed ex-military guys who genuinely appeared rough and homeless.

Summer faded, and the intense heat gave way to rain showers and cooler temperatures. Hot thermal cups of coffee replaced iced bottles of water. I had not been into coffee that much before I started contracting, as for an Englishman, it is always tea. But with the regular dawn runs before the muezzin's call to prayer, waking up at 3 a.m. became a struggle. The CO in the American command centre must have noticed this, as I stumbled in to get the daily radio routine one morning.

'You look half-dead,' he said.

I didn't have an answer.

'Here, take this,' he said, pouring me a coffee. 'It's a cup of joe.'

It was a game changer. A cup of real coffee, not the instant shit I was used to. I felt a jolt of lightning shoot through my body. I was wired for light and sound, and ready to take on the world.

Caffeine … what magical stuff it is.

Gradually, the Americans started securing more and more ground and expanding their area of operations. The list of out-of-bounds routes became shorter. This in turn allowed us to broaden our visits deeper into Ramadi.

This is what our clients wanted. For we were all operating in not only the planet's most dangerous city, but also the most devastated, sorely in need of reconstruction.

12

Death in the River Valleys

Even though the Americans appeared to be regaining some semblance of control in Ramadi, the violence continued unabated. There seemed to be no end in sight, particularly as the insurgents changed tactics. One in particular provided us with even more nightmares, if that was possible.

They started using trucks loaded with explosives as VBIEDs, instead of cars, which meant twice the load and double the devastation. On top of that, they threw chlorine canisters into the mix. In one incident report we discovered that a recent truck bomb had killed thirty, and that a further 350 had seriously damaged eyes and burnt lungs from the gas. Almost all were innocent civilians, including children.

Events evolved rapidly, becoming increasingly gruesome. Captured Westerners were beheaded live on videos which went viral. In the intel reports, the same name kept cropping up: Abu Musab al-Zarqawi. He was a Jordanian who had run training camps in Afghanistan and had become head of al Qaeda in Iraq (AQI), based between Fallujah and Ramadi. He featured in all those early 'snuff' videos and the Americans placed a US$25 million bounty on his head. We all knew that if the guys trying to hit us with IEDs and ambushes ever got hold of us, we would have starring roles in those grisly videos, after having been tortured. Consequently, I made a few changes to my kit and kept loose 9mm pistol rounds in my pockets. If I ever got into a firefight with no escape and ran out of ammo, I would grab one of those loose bullets and shoot myself in the brain. There was no way I was going to allow those guys to parade me in front of the world's media before cutting off my head. Not a chance. I never discussed it, but I am fairly certain that other teammates would have planned to do the same.

Eventually, Zarqawi was tracked to an isolated safe house north of Baghdad, and American jets dropped several 500-pound laser-guided bombs on the building, obliterating it. The most wanted man in the country, whom even his own supporters called 'Sheik of the Slaughterers', was no more.

But that didn't stop the violence, or mean that I would stop keeping loose 9mm bullets handy. Those were still in my kit. Even without Zarqawi on the scene, I was not going to be taken alive.

As we were always outnumbered and outgunned, the plan was never to stand and fight. Instead, we needed to get our clients and ourselves out of harm's way as quickly as we could, so our gunners would lay down a curtain of lead while the drivers pumped accelerators as hard as they could. If any of our vehicles stopped, there was a very real possibility we would be surrounded and overwhelmed. In our world, movement was life … and having my head cut off by a bunch of fanatics was not part of the plan. That's why our driving and shooting skills had to be better than excellent.

Every few weeks we had a 'down day' when we would practise those skills at the shooting range. We regularly included our clients, recreating worst-case scenarios so they would know exactly what was expected of them. At first these were dry runs, but they progressed to live-firing exercises for increased authenticity. Being dragged by a big, heavily armed dude out of a vehicle shattered by an IED, then hurled into another set of wheels and driven off at breakneck speed could be disorientating at the best of times, so we practised it with them time and time again, simulating the heat of battle. If it happened in real life, they would know what to do and hopefully it would save precious time – often the difference between life and death.

The shooting range was across the bridge at Camp Ramadi, the largest military base in Al Anbar province. One day, we had a slot booked from 9 to 11 a.m. This gave SET 20 a chance for a rare morning lie-in. As no clients were attending, we mainly used the time to practise snap shooting.

Just as we were about to pack up, there was a massive explosion. I felt the blast shudder through me like an earthquake and knew it was close. Very close.

At first, I thought it was the main camp gate, but someone else said it was the bridge. Either way, getting back to our camp across the river was going to involve some drama. American soldiers were running about in full kit, clambering into Humvees and revving them in the direction of the gate. The last thing we wanted was to be locked down in Camp Ramadi, so I tagged behind what

I assumed to be the Americans' Quick Reaction Force (QRF).

Swinging out onto Route Mobile and turning right, I could see straight away, by the thick spirals of acrid smoke, that the bridge had been blown. A massive HGV had been detonated by a suicide bomber and the northbound section completely demolished. The wrecked VBIED truck, a thirty-six-tonner, had been hurled headlong into the river by the force of the blast. All civilian traffic had stopped, but we managed to follow the QRF convoy against the counterflow on the remaining carriageway. I assume they thought we were Americans.

An apocalyptic scene of total devastation greeted us. Some bodies were still inside the shattered cars; others had been blown out and were strewn like rag dolls across the road. Hanging out of one car's back window was a man's body so twisted that everything, from his neck to his limbs, was facing in the wrong direction.

As we reached the end of the bridge, a round disc seemingly with spikes on it lay in the road. Driving closer, I realised with a shock that I was looking at the outer skin of a human face attached only to a scalp. What I had thought to be spikes was actually hair sticking up. It looked like a rubber mask.

Nine Iraqis died, and it was obvious why the insurgents had chosen that bridge, even though they had killed and mutilated their own countrymen. Iraq is vertically divided by two large rivers, the Tigris and the Euphrates, and destroying bridges spanning them created supply-chain horrors for the American-led administration. Every blown bridge involved major diversions and route changes, often redirecting traffic straight into the waiting guns and bombs of the enemy. We would usually see at least one or two burnt-out vehicles on these diverted routes, sometimes still smouldering from the night before. Even though night travel was a lethal gamble, I knew many lads who would regularly saddle up in a pick-up, with a machine gun and welded steel plates as their only protection. It was like something from a *Mad Max* movie and, in my opinion, a death wish.

Summer had drawn to a close, with heavy casualties on both sides. But with autumn starting, the situation seemed worse. Most people think Iraq is a steaming hot country, which it is in summer, but winters can be equally brutal. I first learnt of that as a teenager reading Andy McNab's book *Bravo Two Zero*, which had been a bible to me at secondary school. I could easily fool my parents that I was studying, when hidden behind the GCSE maths texts was the iconic story of the SAS patrol in the First Gulf War that lost communication and eventually one another during the raw Iraqi winter. The fact that Iraq had bitter snowstorms was probably one of the few things I did learn in my school years, and as a result, I always packed the best in outdoor gear. For me, Arc'teryx was a pretty good bet for icy weather: solid, reliable, light and easy to pack.

It wasn't just the cold. The changing conditions and gusting winds also generated sandstorms. These were deadly, as visibility was little more than a couple of yards, so there would be no medical evacuation helo if we had a casualty. There is a saying, 'Air cover red, stay in bed', which was advice well worth heeding. It was also always welcome to have a break from risking our lives daily.

Our leave changeover was coming up and I was due home for most of October. Just before we were about to fly out, the SET from Fallujah went on a mission, escorting a client to one of the many reconstruction project sites. Our guys were covering the client when a sniper fired. The first bullet hit team member Darren Bull's weapon, knocking him backwards. Our guys returned fire, moving back swiftly, and peeled off towards the vehicles. Just as Darren was opening the door, the sniper shot him a second time. That time it was fatal.

Darren was our first loss in the Al Anbar area of operations and it was devastating. The Fallujah guys were a cracking bunch, extremely professional, and I had known many of them from the beginning of the contract. We were really tight, like family, and Darren was among my favourite people. Seeing the team heartbroken was deeply distressing. What was also really sad was that Bully – as we called him – was on the same leave rotation as me and eagerly looking forward to taking his children away on a Center Parcs holiday. The world can be a shit place.

We picked up the Fallujah guys on rotation leave, minus Darren, for the run to Baghdad airport. There was silence for most of the trip.

I spent the festive season of 2006/7 with Nicky and I'm the first to admit I wasn't much of a boyfriend at the time, always looking for the next buzz after getting home. Brawling and drinking topped the list. I was a nightmare to live with and it only got worse over the boozy festive period. Nicky and I welcomed in 2007, but my mind was only focused on getting back to Iraq. The intense fighting was constantly on the news and the lads messaged daily, telling me what was going on. I needed to be with them.

Leave was soon over and, once again, I headed back to another twelve weeks of adrenaline rush. I loved being with Nicky and it was nice to have a family like hers caring about me, but the siren call of action and adventure at that stage of my life was too strong. I struggled to shake it.

It was snowing in Ramadi when I returned – not a lot, but enough for it to be a bloody nuisance. The triangle of death had transformed into a slippery mud bath, and thick clay smothered our kit, particularly boots, clinging like dog crap and making them sticky and cumbersome.

Attacks on us were also becoming more frequent – averaging one a week – although I barely counted the number of times we were struck by gunfire, as it was usually shoot and scoot. IEDs were different and far more deadly. Those, I definitely counted; I survived thirteen. For me, a lucky number.

The first attack after I got back was during a return trip from Fallujah, when we were stopped at a roadblock. The Americans had found a large daisy-chained IED on the side of the road – a string of bombs wired together to go off at the same time – and Route Mobile was closed. We sat stationary in the traffic for a while: the most vulnerable situation imaginable for us, with a client on board. Everyone hated being static and Daryl was looking around for alternatives. Off to the right-hand side of the main road was a track used by locals. It was sand and easy to hide an IED, but then again, sitting in a car where a daisy-chain of IEDs had just been discovered was not exactly reassuring either.

Daryl broke the silence. 'Everyone good for a push, right?'

He was basically thinking out loud. One of the finest leaders I have ever had the privilege to serve under, he would always consult the team in dodgy situations like this. He genuinely valued everyone's input, although we always knew the final decision was his.

'Oh yeah,' said Shaun. The team's second in command was in vehicle three. His lively East-End accent, filled with cheekiness, as well as that irrepressible cockney humour, always cheered me up. He reminded me of Buck, the weasel, in the *Ice Age* films – a proper crazy cat. But he was also a highly experienced soldier who had seen a lot of action in Northern Ireland. Like Daryl, he was one of the finest natural-born leaders I knew.

Daryl nodded at me and I drove off Route Mobile, pulling in behind a line of battered orange-and-white Iraqi taxis and even more battered pick-ups. As we knew the main road was the IED target, we hoped the sandy bypass would be okay. I could actually see the bomb-disposal robot moving around the tarmac to the left of us, like something from *Star Wars*, as the EOD (explosive ordinance disposal) guy in his green armoured suit remotely controlled it. As we drew level with them, a huge explosion went off behind us, shattering eardrums and filling the car with stinging sand.

'Contact! Vehicle two!'

This was the worst possible news for us. Our client car had been hit directly by an IED. Daryl and I looked at each other, waiting for an update.

'Two is good, client's good. Drive, drive, drive!'

'Three?'

'Three is good and moving.' Shaun's cockney voice was calm and reassuring. We were all okay, despite the second car having borne the brunt of the blast. But the armour had done its job, protecting the client inside. Our relief was palpable.

However, vehicle two's steering was stuffed beyond redemption. Somehow the driver managed to coax it limping back onto the main road. Fuel gushed from the ruptured tank and we all knew there was no way it would get back to camp under its own steam.

Shaun rushed in to assist. 'Three is in to push.'

We had to shunt it away from the contact area. Immediately, in case more IEDs went off. Shaun's driver rammed the metal fender of the gun truck against the back of the client car and started revving. I could see bits of bodywork flapping off like metal confetti, as vehicle two gained momentum. The tyres had been ripped off and the vehicle skidded about on bare rims, as the driver frantically tried to control it. It was like a *Candid Camera* clip – metal, fuel and sparks flying everywhere, with the rest of us laughing like crazy as the crippled vehicle slewed down the road. There was the serious risk of a spark igniting the fuel, but the bigger danger was to hang about.

We reached a large sand berm about a mile further down, which provided us with some cover. The team jumped out and, as gently as possible, manhandled the client into the back of our car. Vehicle one had become the client car, while Shaun's crew tossed a towline to vehicle two. It was absolute chaos, but a chaos that we had rehearsed many times. We thrived on it. All those training sessions at the range proved their worth a thousandfold.

Daryl gave the client a once-over physical check as I sped out of the contact area. The client, our top priority, was okay. The pace was furious, but in those life-or-death situations, time always seemed to stand still. No matter how fast you are going, everything appears to happen in slow motion. We dragged the client car back into camp on its rims, as people watched in total disbelief. It looked as if it had been through an industrial shredder. Yet, only one single piece of frag had pierced the armour, ending up lodged in a box of A4 paper.

A later assessment indicated we had been hit by a classic secondary device, as the main one on the road – the daisy-chain – turned out to be a fake. It was purely a distraction technique, and it certainly worked, as the American EOD squad had been entirely focused on neutralising a hoax just as we got blasted. The guys in vehicle three said that at least four nearby civilian cars had been destroyed and there had been almost certainly some fatalities.

But perhaps one of our weirdest attacks actually started in the dentist's waiting room. I had a painful tooth that had been bugging me for a while and, as I had access to most US privileges at the camp, I visited the American dentist.

'Grab a seat, sir. Waiting time is an hour or so, and the coffee is there.'

I picked up some American soldier magazine and an article that caught my eye was about a new insurgents' tactic. Instead of sniping from buildings, these guys were lying in the boot of an estate car and, having removed the brake-light cover, fired out of the gap. All car lights in Iraq were smashed, so a broken brake light was actually the norm. There were no traffic cops around, in any event.

A couple of weeks later we were stuck in an afternoon sandstorm at Baghdad airport and couldn't do a 'run for the sun'. Instead, we opted to stay over with our client and get back to Camp Blue Diamond first thing the next morning. Whenever we did this run, the gun truck would lead the way out and hold the traffic merging onto Route Mobile, so that our vehicles remained in a tight group. We could not be separated, for obvious reasons. With Daryl on leave, I was commanding the gun truck and a fast-moving car came up Route Mobile just as I blocked a lane for our convoy to filter in behind me. However, instead of moving away, it ended up behind, as an approaching American unit coming up the road also wanted it out of their way. As the Americans turned into the junction that we were about to turn out of, this particular vehicle was caught between us.

That's when I noticed it – the rear-left brake light was missing. Exactly as I had read about in the dentist's waiting room.

I called out to Will, the gunner. 'Mate, keep eyes on that car.'

I then got on the radio to the rest of the team and let them know this vehicle was a red flag. Watching it in my side mirror, I pulled my M4 rifle out from its position wedged between my seat and the centre console.

'That car is two up, H, but one just turned and looked behind him,' said Will. 'Someone is in the back.'

'Knew it, mate,' I replied.

We could see two people in the front of the car, but one had turned around to speak to someone we couldn't see. That meant a third person was in the back of the beat-up estate. His mate's body language had given him away.

Still watching in the side mirror, I saw the driver's hand slide off the steering wheel onto the door handle.

'Standby lads,' I said. 'He's going to move on us.'

In what felt like slow motion, the driver opened the door. I could then see his AK-47 as clear as daylight, as he raised it, pointing directly at the back of our gun truck.

'Fire!' I shouted.

Our lives depended on that call. A second's hesitation would have been too long. The air filled with cracks like snapping branches, as my gunner put down a burst of 5.56mm into the insurgent's chest. The rounds went straight through him and into the guy in the passenger seat. He slumped forward, head on his chest. Both men were no longer threats.

I then noticed the car's suspension shift. The hidden third guy was moving about inside, ready to come and join the fight. We waited for him.

'Another burst, H?' Will asked.

I had to give permission to fire and at that moment I didn't know who or what was in the back. It could be a sniper; it could be children. Company rules were that we were only allowed to protect ourselves if an immediate threat to life was imminent. Such as what had just happened – a man getting out of a car and pointing an AK-47 at us. But I didn't know what would happen next.

'Hold, mate,' I said.

The back door nearest to the gun truck opened.

'H?' my gunner asked.

I replied instantly: 'HOLD!'

Then I saw him. A head popped out. He was just a boy, small ... definitely no older than sixteen. Maybe as young as twelve. He was trying to climb out of the car while dragging a Dragunov sniper rifle almost as long as he was. As he exited, the rifle fell to the floor. Unless he had a drop weapon, we could not engage.

I levelled my M4 rifle at him. We locked eyes. He knew it was game over and turned to sprint into a grassy area behind us. Tracking him with the red dot of my EOTech sight and finger on the trigger, I waited for him to draw a secondary weapon. He ran up a hill to a cluster of buildings and then turned, looking directly at me. If he was going to shoot, this was the time for it.

'Don't you fucking do it,' I whispered.

He disappeared between the buildings and I told Will to keep watch as I climbed back into the vehicle.

That day I didn't have to shoot a kid. Would he ever know how lucky he was? Maybe, maybe not. But hopefully he had a change of heart and reconsidered his line of work.

Would I have been able to live with a clean conscience if I had killed him?

Thankfully, that's a question I didn't have to answer.

13

Farewell to a Brother

A few months later, we had to pick up an IT client who would be staying for a few days at Camp Blue Diamond while working on a project. I was back to driving vehicle one, as Daryl had returned from leave, and Shaun was commanding vehicle three, the back-gun truck.

That suited me fine – we were back to normal. The A-Team. In those days, there were few things more reassuring than having Daryl and Shaun in charge.

The next morning, we headed off to HQ Ops in Fallujah with the client, where he would be working that day. It was the usual routine: Daryl went in to talk to management while we chatted to the Fallujah lads. When he returned, he was raging. Apparently, we were supposed to take the IT guy back to the airport. A few days with us turned out to be one day, so we had to do two 'runs for the sun' in succession, always a dicey option. No one had informed Daryl of this and, as a result, he had a bust-up with ops staff. It escalated into a heated argument and eventually Daryl stormed out. He was usually a level-headed Jock, but fiery as they came and standing his ground when he was in the right.

By the time we got back to Blue Diamond, reports of the altercation had gone right up the corporate chain to the top of the company and, as a result, Daryl was told he would be moving teams. I was completely gutted. He had been our leader from the beginning and to lose him over some bullshit spat was even worse. God only knew who they would send to replace him.

The team was on a real downer, as Daryl was extremely popular. We had been through so much together, watching each other's back in numerous roadside attacks, and to see him go hit me hard. I had learnt so much from him in those first years in Ramadi, lessons that would always stand me in good stead. We said our goodbyes: basically, little more than a bear hug and backslap, and a muttered, 'See you soon, mate.' Just the usual bloke shit. It's a day I will never forget. It's also a day I will regret for the rest of my life.

I wish I had said more to him. I wish I had been open and upfront about what he really meant to me. I wish I had told him what an inspirational guy and superb leader he was. I wish I had told him that he was the best.

But I fucking well didn't.

On 12 June 2008, Daryl was hit by an IED and killed instantly. I was working out in the gym and one of our lads came to call me. I knew something terrible had happened and we walked back to the ops room in silence. The news was broken to us and my heart sank. There are things in this world which cannot be put into words, and this was one of them.

Maybe one day I will meet him again, in some other place, and try to say the things I should have said. Maybe he already knew, and words were not needed. But even so,

I should have said them. Since then, I have never left good things unsaid, if I could help it. You never know if you will get another chance.

The whole team could not attend Daryl's funeral, as we had to remain operational – something he would have totally understood. I was one who would have to stay behind. Shaun would be going, as it would have been unthinkable for him not to be there. He and Daryl had backed one another up through thick and thin, making tough, educated, tactical decisions in the lethal ticking time bomb that was the Ramadi snake-pit at the time. I think most of the team agree that it is largely thanks to Daryl and Shaun that we are alive and with limbs attached today.

The funeral took place in Scotland. Daryl's wife Tracie played host to the private contractors turning up the night before the funeral to console each other and pay their respects to Daryl's family. But also – more importantly – to celebrate a remarkable life.

Daryl's youngest son Ciaran was only seven at the time, but he remembers that night vividly. Fourteen years later, when both of us were attending Shaun's wedding, Ciaran told me about it. He said his dad's casket was in the front room, and there was a lot of noise outside.

'It was early in the morning, before sunrise. The garden was full of strange naked men covered in tattoos, cheering and laughing as they threw one another down my tower slide. I had no clue who they were, but I knew I wanted to be one of them.'

I understood exactly what Ciaran meant. Shaun and the boys were saying goodbye to Daryl in their own wild and unique way.

Ciaran and I drank into the early hours of Shaun's wedding, recalling our days with Daryl. It was a joyful night and, towards the end, Ciaran quipped that all his dad had left him with was 'emotional issues and a receding hairline'. We fell about laughing in the hotel bar, and for the briefest of seconds, I thought it was Daryl speaking. That black, quick-witted humour could only have come from his mouth. Daryl was right next to me – it hit me like a sledgehammer how deeply I still missed him after all those years. The harsh lump in my throat was softened by laughing with his son. This fine lad was a diamond chip off the proverbial block.

The Ramadi team limped on without Daryl for a while, until we got an older guy called Stevie Orr, an ex-Irish Guards veteran with the same 'take charge' attitude as Daryl and who also smoked like a trooper. He had seen some serious shit in the Northern Ireland Troubles and I liked him right away. I'm not sure if that was instantly reciprocated, as I had gone a little rogue after Daryl's death and returned for duty sporting a black eye.

My new team leader was not impressed. In his crazy Northern Irish brogue, he asked what had happened.

'Just scrapping,' I said.

'You need yourself a hobby, wee man.'

Stevie was a passionate biker and over the next few weeks talked me into buying a motorbike to occupy my downtime, almost twisting my arm to book a test. Once I had done that, I wasn't sure which bike to go for. Stevie had that covered.

'Come here,' he said, flicking open his laptop. On the screen was a sleek tornado-red thumper. This was as far removed from a conventional road bike as a poodle from a wolf.

'It's a Triumph Daytona 675. Deposit's paid. It's in Aylesbury and we pick it up next week.'

Stevie had the same leave period as I did and lived in London. It's the best thing anyone could have done for me at that time. I don't know how he knew, but with Daryl dead, he somehow sensed I was in need of direction. A bike was the perfect distraction.

'It's like this, mate,' he said as we drove to fetch it. 'You put the helmet on, start the bike and you're away. Nobody can talk to you and you don't have to say anything. Just you and the machine. No bullshit, no talking, no drama. Just ride the fucking thing and it will change your life.'

He was right. That was when I fell in love with motorbikes.

I had no gear and Stevie gave me a helmet, leather top and bottoms, and boots. The first run we did was to Clacton-on-Sea. My God – it was like being on a rocket! I had ridden a Honda 500cc upright for my driving test, but this machine was a totally different breed. To be fair, my instructor had warned me about that when I told him I was getting a Daytona. The power was instant. Like magic. The slightest twist on the throttle hurtled me forward, damn near ripping off my arms. I quickly learnt to lean into acceleration and put my weight onto the tank, stopping the front end going light and lifting like a plane.

I spent that magnificent day riding with Stevie and slept over at his place. I could not have been happier – my first official day as a bike owner. After a few drinks, I went to bed with a huge smile on my face.

The next morning, we rode to the motorway and pulled up on the bridge. Stevie told me to take it steady, then watched me go as I slid off the exit road to Lincoln, like a baby bird flying from the nest.

Twenty minutes later the heavens opened. Torrential rain bounced off my visor, making visibility almost as bad as an Iraqi sandstorm. I contemplated pulling over, but I really wanted to get home. Soon I was soaked to the bone doing seventy miles an hour, dodging HGVs sluicing rain like shrapnel and a freezing wind that cut to the marrow.

By the time I reached Lincoln, I was so numbingly cold I could not feel a thing. Kicking the side stand down, I left my new machine in the deluge and trudged into the house.

'How's the new bike?' Nicky asked with a wicked grin, as I pulled off my wet kit, puddles pooling on the kitchen tiles. A hot bath sorted me out and as I lay in the steaming water, I excitedly told her about my new passion.

She listened, smiled and rolled her eyes.

'Boys and their toys.'

That bike helped me to get back on track, and to this day I'm a massive fan. There is something magical about riding off, not knowing or caring where you're headed, just enjoying being at one with the machine. It was a changing point in my life. For many people, big powerful bikes are a symbol of reckless wild youth. For me it was the opposite – getting my life back on track.

That leave rotation with my new-found passion went all too quickly, and soon I was back in Iraq. It was late 2008 and a fresh battalion of US Marines had arrived in the country. Clean-faced youngsters took over from the current battle-scarred and exhausted fighters, who looked a decade older than their replacements.

The newbies got the most terrifying welcome possible – a truck suicide bomber. It was early in the morning of 22 April 2008 and two youngsters stood post at the main gate of the Joint Security Station as the suicide truck weaved through the chicane barriers. Everyone there, including Iraqi policemen, ran off as they knew what was about to happen. But Corporal Jonathan Yale and Lance Corporal Jordan Haerter stood fast, hammering the truck with automatic fire. They killed the driver, but the suicide bomber had a dead man's switch that detonated the moment the spring-operated device was released. The truck exploded at the gate, killing one of the brave Marines instantly, and the other a few hours later. These two men saved the lives of at least fifty colleagues. For that supreme act of heroism, they were awarded the Navy Cross, one level below the Medal of Honour. I saw the CCTV footage and I salute those two young warriors.

Our client's latest project was building the Al Anbar headquarters for the newly formed Iraqi police force, not far from Camp Ramadi. The country was about to be handed over to a soon-to-be democratically elected government and security sites for the incoming administration were springing up all over the place.

I hated that site. It was exposed, with cars parked everywhere and lots of random Iraqis with no security clearance wandering around the premises. Our client regularly wanted to go there, and we had to take him, despite it being a security nightmare. I was always on edge. Within a few months the site had been completed, all paid for by the US government, and a grand opening planned. Naturally, our client wanted to be a part of all the pomp and ceremony, as it had been his company's pet project.

Stevie immediately put a stop to that. As a veteran operator, he didn't mess about when it came to security catastrophes like this. He flatly refused to take our client to the ceremony. A much-publicised grand opening in

Ramadi was like painting a bullseye target on your forehead with a 'just fucking shoot me' sign. It caused huge drama with the client, but Stevie stood his ground.

Thank God he did. It turned out that the builders of the client's much-vaunted project were actually insurgents and had packed the walls of the building with explosives. Once the local police chiefs and other administrative bigwigs were inside, admiring the brand-new complex, some jihadist pushed the detonator. The entire building went up in flames and rubble, killing everyone inside.

I reckon after that our clients figured that whatever they were paying us, it wasn't enough.

By then, I had been in Iraq for four years and things didn't seem to be changing much. Most routes were still classed as 'black' after sunset, so each night I hit the gym for a training session. I loved working out, but this was more of a survival programme than anything else. You could not operate effectively under the extreme triangle-of-death conditions if you weren't fit. We carried heavy kit, wearing Nomex suits and body armour in the summer desert heat and freezing winters, and to call this merely 'energy-sapping' is seriously understating the case. Being flabby meant you were not only a liability to yourself, but to the entire team who would have to drag your ass out of the kill zone. This was no country for soft men. I despised guys who came here wheezing out of their asses and endangering the rest of us. The saying, 'If you can't keep up, don't step up' was the first commandment as a contractor in Iraq.

I knew a few lads who came into the country, got hit by an IED and caught the next plane out, and there's no shame in that. If fact, I admired them for admitting that this was not for them. But for those out of shape, I had no sympathy.

However, from 2009 onwards, I noticed a real change in the private-security world, and sadly not for the better. The need for governments to contract out security obligations in war zones had by then increased beyond all recognition. The demand for contractors rocketed, with the result that once rigorous checks implemented by security companies were significantly reduced.

It affected us directly. One individual who came to Ramadi was an instant red flag to me. We first placed him as a driver, but he wasn't happy with that, constantly complaining that he wanted to be a gunner. No way did I want that, and eventually the guys at Fallujah did us a favour and took him from us. He did not last long there, either.

His name is Danny Fitzsimmons, and luckily for us he left our company, getting a job with another outfit operating out of Baghdad's Green Zone. One night, while playing cards, he killed two colleagues and injured an Iraqi security guard. He claims it was self-defence in a drunken brawl, but the judge did not agree. At the time of writing he is serving out the rest of his twenty-year sentence for manslaughter in a British jail.

It was a dark day for the security world. Many of us then knew that we needed to be careful about who was guarding our backs.

At about that time, our team moved from Ramadi to the Al Asad Airfield, two hours' drive away. We had frequently been to it on tasks before, so knew it well but had never been based there. Its most notable feature was the number of rusting planes from the old Iraqi Air Force dumped outside the camp. Apparently, someone had once climbed into a cockpit of one for a photoshoot and accidentally set off the ejector seat. I don't know how true that story was, but it would have made a great selfie.

It was tough leaving all the great people I knew at Blue Diamond and Camp Ramadi. I had spent five years of my life there and loved it: the adrenaline, fun and the experiences – some tragic – would be with me forever.

That would not be the case with Al Asad. For a start, it was massive and our accommodation was on the opposite side of everything else. Just to go for a meal or to the gym was a twenty-minute drive. Also, it seemed the people there did not want us at the camp. The friendly banter and brotherhood we had with the American Marines and SEALs at Ramadi did not exist there. I know a lot of people resented contractors for the money we earned and I understood that. So we kept a low profile.

In short, Al Asad marked the beginning of the end for me in Iraq.

With Nicky by then having endured five years of me being consistently away with saint-like patience, it was time to consider going home. She wanted a family and a stable life. I often joked she was a rubbish 'trophy girlfriend' because she wasn't into material things. She just wanted to be with me and, as I was increasingly realising, I wanted to be with her just as much.

Despite that, she never asked me to stop doing what I did.

In 2010, I handed in my month's notice and decided to try my hand as an 'adult' in the big world. It was gutting to say goodbye to close friends, and some said I would be back. To a certain extent they were right. I would be back, but not in Iraq.

My transition into civilian life resulted in getting a qualification as a personal trainer. It was an easy enough job and I loved helping people feel better about themselves, watching their confidence grow as they became leaner, stronger and fitter.

We moved to a spacious, four-bedroom house just up the road from our first home. Nicky was happy enough and we lived a normal life there for a few months. But she was always fatigued, sleeping on the sofa for a few hours every day.

We soon discovered why. A certain little person was about to rock our world.

14

Somali Pirates

Alexa Jasmine Collins was born on 12 June 2011, a beautiful baby girl, who would have the deepest influence on me.

It was a traumatic birth and Nicky lost a lot of blood. In fact, it was the most blood I have ever seen, so knew first hand how touch-and-go it was. She had to have a number of transfusions and was confined to her hospital bed for several days to regain her strength.

We did not know the sex until the baby was born, so had only bought gender-neutral clothes and toys for Lexi. As Nicky was still recuperating, I left her and Lexi to go out and buy some dresses, ready for when they were allowed to leave.

I hadn't been out of the hospital building for three days and felt as if I was dreaming when I walked into Mamas & Papas, a shop I would never in my wildest dreams have thought of visiting in my past life. I found the newborn section and grabbed everything that I liked, crossing fingers that Nicky would as well. If not, at least she had a few options to choose from. It was the first time I had ever bought clothes for anyone so small, and it rammed home the reality that this little baby was my responsibility. I would have to clothe her, feed her and keep her safe. The reality of fatherhood was rapidly sinking in, and I liked the feeling.

I knew without any shadow of doubt that Nicky would be an amazing mother. She was born for the job. However, I probably needed to have a good hard look at myself – and I did just that. I silently vowed I was not going to be a bad dad, as I had seen first hand in my own childhood how *not* to do parenting. This was my chance to raise a good human being and put something into this shitty world that would shine bright in the darkness. The symbolism of her being born on the same date that Daryl had died has never been lost on me. She was a gift, and I had to do it right. I could not waste this opportunity.

For the next few months, we nested in our new house. I was out early and back late with my job, and Lexi would often be asleep when I left or when I returned. I hated that, and sometimes I would wake her up just to see her gorgeous eyes. She disliked being on her own. If Nicky or I tried to sneak out of her bedroom, she knew it instantly, crying until we picked her up. We gave up in the end and she slept in the cot beside us. Eventually, this progressed to moving her to the middle of the bed and she slept with us for a good few months. I loved waking up to see her little face first thing in the morning, listening to her chatting away in baby talk.

She enjoyed her sleep, just like her mum, and would go absolutely ballistic if we woke her after she had nodded off in the car. Many a time we found ourselves sitting mutely as she slept on the back seat, not daring to disturb her. It just wasn't worth it. I thought working in Iraq was dangerous, but that was nothing compared to rousing a sleeping Lexi.

She grew to be a smart toddler and was quick to learn from her environment – a curious little girl with a personality to match. A mix of a little of me and a lot of Nicky.

I continued working as a personal trainer for a while longer, but the itch to get back into the excitement of private security was always there. However, fatherhood was keeping me grounded, as I loved being a dad. It was just the price one had to pay.

That all changed when I had a run-in with my bosses. My motorbike broke down one morning and, without transport and needing to get it repaired, I missed a full day's work. However, I worked longer hours over a weekend to make up for the lost time, but they still weren't happy. Consequently, company management called to say that they wanted to 'have a chat' with me the following Friday.

One thing I have never taken well is my actions being called into question, and an insinuation that I was a liar was even worse. I took two days off to figure out my next move. I knew I had other options but needed to see what was available. Luckily, a friend came through for me, and I almost instantly got a job offer. I accepted it.

However, it happened so quickly that I hadn't told my bosses about my decision. The next Friday, the day they wanted to have their little 'chat' with me, my phone rang. It was the manager.

'We're here for the meeting,' he said. 'Where are you?'

'Heathrow.'

There was a confused silence.

'Heathrow Airport?'

'Correct.'

'What?' I could hear the incredulity in his voice. 'Why?' 'Flying out for a new job. Hope your meeting goes well.' I hung up.

I had lasted longer than I had thought I would in a normal job. The plus side was that I got to watch Lexi grow up from being a baby to a toddler, to hear her say 'mama' and 'dada' for the first time, and to see her take her first steps. That was priceless: the kinds of things that really do matter in life. Gold dust, for which I will forever be grateful.

My new job was in maritime security, preventing large ships from being hijacked off the lawless coast of Somalia by well-armed pirates in speeding skiffs. That's why I was at Heathrow and about to board a plane to Cairo when my now ex-manager called.

Once in Egypt, I picked up some weapon cases, body armour and radios, and boarded a cargo ship at the Suez Canal. From there, I would 'ride shotgun' as a four-man security team along the IRTC (Internationally Recommended Transit Corridor) down the Red Sea into the Gulf of Aden, past Somalia and across the Indian Ocean to Sri Lanka.

We would disembark in Colombo and wait for a ship returning from Sri Lanka to Suez through pirate-infested waters. Somali gunmen had hijacked a number of cargo vessels and oil tankers, causing maritime panic and disruption to global trade. The Gulf of Aden was considered the most hazardous shipping area in the world. The pirates' methods were simple, but effective. Armed with AK-47s and RPGs, they would chase down ships in fast skiffs powered by outboard motors, clamber up crude ladders that they hurled up the sides like grappling irons and hold the crew for ransom. Once that was paid, they sold off the oil in the tankers and the vessels back to the original owners. It was big business, raking in millions of dollars.

Our four-man team would do six hours on and eighteen off, day and night. If we spotted any fast-moving speedboats or skiffs, we would position the remainder of the team on the bridge in a 360-degree all-round defence. At night we were at our most vulnerable, relying heavily on radar, unless there was no cloud and a bright moon. We could then spot the luminescent propeller wash of any approaching craft.

I had several encounters with Somali pirates, but nine times out of ten, just holding my weapon above my head to show that we were armed and dangerous was enough to deter them. If not, we fired warning shots. If they still didn't turn around, we would shoot and cripple the engine of their boat. Killing the would-be attackers was the absolute last resort, as many of these pirates were just out-of-work fishermen. Fortunately, we never had to resort to that. Most pirates preferred to avoid guys like us, who were prepared to do battle.

With tropical weather and bars stretching all the way along the coastline, Sri Lanka was a great place to grab a few days off. We would sunbathe and swim in the warm waters of the Indian Ocean, then have beach barbecues and beers in the evening around blazing log fires. I loved sitting there with an ice-cold beer, listening to waves crashing on the shore and having a laugh with my teammates. And getting paid good money for it.

I even found a local who owned a Suzuki RM250 motocross bike. Most of the hire shops had only mopeds, but I wanted to go a bit further afield than the seaside resorts or beaches. The ship that we would be guarding on its way back across pirate-infested waters was still four days' sail away, so I had plenty of time to explore the island.

After a bit of haggling, the owner of the motocross bike rented it to me for a few days, and I packed a bag and set off, earphones playing my favourite biker playlist, with songs ranging from AC/DC and Whitesnake to Tracy Chapman and Everything but the Girl. It was total freedom: music in my ears, wind in my hair, sun on my face. Well ... until the daily tropical storm burst, that is. It happened at 3 p.m. every afternoon and you could almost set your clock by it. For about thirty minutes the clouds would open and I would pull off the road. Then, as suddenly as it had arrived, the rain stopped and the sun re-emerged, hot and brilliant, and I would set off again.

I followed the sweeping coastline all the way down the island. The locals were crazy drivers, ignoring almost all traffic rules, and I needed eyes in the back of my head. Not really a massive problem for someone who had spent the past five years driving in Iraq's triangle of death.

It felt good to be that free. I would ride until I got hungry, have something to eat and then, after the daily tropical storm that cleansed the island had passed, I would ride again until I found a beach bar. These opened and shut when they wanted to. Closing time was usually when the last customer finished drinking. Once he left, the barman would shutter the place and sleep on the floor until the day-shift guy arrived to open. I was that last customer on most nights, and when I was done, most bars also had a wooden beach hut with a hammock nearby, which I could sleep in for a small price. I pushed the bike inside and would fall asleep to the sound of the waves.

That was one of the best and most carefree adventures I've had. I sometimes went all day without saying a word to anyone, and if I did speak, it was to order a glass of wine or a cold beer. Sri Lanka is not far off the equator and after a day in the beating heat, nothing felt better than a solitary frosted beer on a tropical beach, watching the sun setting in vivid shades of crimson.

During my time as a maritime guard, I did a number of jobs on various different ships, some huge and some small. I even guarded two tug-boats sailing across the Indian Ocean, with another lad who had been with me at 40 Commando. It was almost relaxing after the constant hatred and violence in Iraq, even though we were never sure of what was behind the next radar scan of the ships we were guarding.

However, about a year into the job, I had a freak injury. A lot of contractors boxed, and we would bring pads and gloves to train to keep fit while on board. I had always kept up with my boxing, as I enjoyed the discipline and concentration that went with it. On one occasion, a lad and I were training on deck and doing body shots on pads, a drill I have done thousands of times. I was working well when suddenly I heard a 'ping' on the inside of my left elbow and my arm buckled, going completely dead.

The pain wasn't bad, but it felt weird. When I looked down, it got even weirder – where my bicep was meant to be was now loose skin on bone. Instead, there was a massive lump on my shoulder. It was the bicep muscle. It had snapped and retracted upwards into my shoulder.

With a virtually crippled arm, I managed to see that job to the end, and went to a doctor once back home. He confirmed it was a snapped bicep and booked me in for surgery at a private clinic in Chester. The procedure proved to be more problematic than the surgeon had originally thought, as they could not grip the bicep tendon firmly enough to pull it down. So they had to slice across my forearm to pin it. As the tendon had lost its stretch, this too proved extremely difficult.

I went under anaesthetic that morning and when I woke, I was in pitch darkness. Confused, I wasn't sure what was going on.

'I'm going to be sick.'

I said it to myself, but someone switched the lights on and replied, 'It's okay, lovey.'

The night nurse handed me a sick bucket and I duly threw up. I had been under for ten hours and pumped full of drugs to keep me asleep. My arm was strapped up to my neck in a black sling that stopped it from bending back at the elbow. I gradually had to click it open one notch at a time, slowly stretching the tendon and eventually allowing myself to lower the arm to my side. Once it completely straightened, I then had to rebuild its strength.

It was like learning to walk again. I could barely bicep-curl a tin of beans and the arm took a long time to build up.

While off injured, I applied for a few jobs to see what was out there. Maritime security had been fun but after a few years of living out of a bag, it had started to become tiresome.

It was time to move on.

15

Back in Kabul

While in rehab, nursing my snapped bicep, a guy I had worked with in Iraq rang and asked if I was interested in a job with a company that had a security contract with the Embassy of Japan in Afghanistan. He knew I was in the market for work, and he said that this was tailor-made for someone with my skills and experience.

The pay was good and in US dollars, which was a bonus, as the exchange rate was favourable at the time. Shift rotation was eight weeks on and four off, so Nicky was okay with the set-up, as I would be returning home every two months. She was working part-time at the local council, and Lexi was happily settled in school.

I took the job, but once again, it meant saying goodbye to my wife and daughter. That never got easier.

'It's all good. I'll be back soon,' I said as we hugged tightly.

Landing in Kabul Airport was a shock. I had last been there in 2002, when the runaway had been littered with wrecked Russian planes from the Soviet invasion and the terminal was a decrepit, run-down building. No more. The gutted aircraft remains had been bulldozed away and the terminal spruced up into a fully functioning commercial airport.

Kabul itself had also changed beyond recognition. The city centre had become a thriving metropolis where music blasted from cars, and people with trendy haircuts, wearing designer-label jeans and T-shirts, chatted away on the latest smartphones. Most women had discarded their burkas, opting for headscarves and fashionable dresses, and they walked freely on the streets, without male minders. It was light years removed from the almost medieval tribal land I had arrived in as a Royal Marine twelve years ago – so much so that I could barely find a single reference point.

I soaked it up, marvelling at what had happened in such a short time, as I headed to the Green Zone. This was a massive, heavily fortified area surrounded by concrete blast-buffering T-walls, slap-bang in the middle of a city of 4.6 million people and a stark reminder that despite all the modernisation, Kabul was still at war. The Green Zone fortress housed most of the foreign embassies, as well as the residences of politicians and VIPs, and the palace of President Ashraf Ghani.

I was dropped off at one of the checkpoint gates, where dogs sniffed my bags and a heavily armed security guard frisked me, before radioing through to the embassy. Eventually, a big dude called Mitch arrived to fetch me. He was an ex-Marine and a keen rower, which I could tell just by looking at his powerful frame.

We walked to the embassy complex and through the compound. To the left was a row of garages housing white Toyota pick-up trucks – all B6 armour apart from one that belonged to the Japanese ambassador. That was a B7, capable of withstanding armour-piercing ammunition.

Mitch then showed me around the operations room, which had maps pinned on walls and computers blinking on desks. In the corner sat the watch-keeper – the link between security contractors on the ground and the ops room. If anything went south, he was the one who relayed that information. In a supremely dangerous city where every diplomat, every politician and every Westerner was a prized target, this was a critical job.

I was given body armour, a radio, a mobile phone, an AK-47 and a Glock 9mm pistol, which I would have to sign for each day. I was then shown to my room. It was small, but had its own bathroom, although I was told that the roof leaked and I would need to move out before the rainy season arrived. Awesome.

I went downstairs to meet the team I would be working with. They came from many different branches of the military, and some I already knew or had heard about. Others I was meeting for the first time. One of them handed me a mug of tea and we started chatting outside the small kitchen. This was a subtle testing ground for both them and me – sort of sniffing each other out, so to speak.

The questions seemed ordinary but the information was crucial: where did you train? Where do you live? Do you know so-and-so? Whereabouts were you based? What are your skills?

In other words, it was an informal 'job interview' for them and me, but absolutely essential, as we would be trusting each other with our lives. This was no place for bullshitters.

If the answers satisfy everyone, trust is established. Luckily for me, my Iraq contracts had been very high profile, so they had heard of me. I was quickly accepted.

I would be working as an IBG (individual bodyguard) for the Japanese diplomats. Two IBGs would always accompany a diplomat in two cars, so there was always back-up. If we were escorting more than one diplomat, we would take an extra IBG, but usually we operated in pairs with local drivers. As a result, I was quick to discover which drivers were the best and made a mental note to be sure of having them on my tasks. However, this was not always possible, and every now and then I ended up with one guy who would make me instantly carsick. He was horrific – stamping on the gas or the brakes, the car lurching forwards, backwards or skidding sideways across the road. I sometimes thought that if the Taliban didn't kill us, he would.

I also started using the gym again and began slowly getting strength back into my arm. An 8.8-pound bicep curl was all I could manage, but it was enough for now. At least I was no longer struggling to hold a tin of beans! Bicep curling with a light weight in one hand and a 26.4-pound weight in the other was an odd experience, but getting strength back in my arm made me appreciate that the extensive rehab after surgery had been worth it.

Kabul reality hit hard – and quickly. Within a few days of my arrival, there was a massacre at one of the city's most famous hotels. The Kabul Serena, very popular with Westerners and diplomats, is in the middle of two of the busiest city roundabouts, with heavy traffic streaming past on all sides. On 20 March 2014, during a function celebrating the Persian New Year, teenage Taliban militants smuggled pistols inside their socks and somehow got past the hotel's security checkpoint. They hid in the toilets for three hours, and once the venue was full, they stormed out, murdering nine people, including small children and women. Afghan special forces surrounded the building, eventually killing all four gunmen.

I had driven past the hotel the day before, and it was a sobering reminder that despite the massive security, we were operating in one of the deadliest areas in the world. As IBGs, we travelled throughout the city every day, visiting many high-risk sites while escorting people who topped the Taliban kill list. Incessant vigilance was the difference between life and death.

I soon settled into the security-circuit routine. Among my favourite politicians was General Abdul Rashid Dostum. He was the founder and leader of the *Junbish-i-Milli Islami Afghanistan* (The National Islamic Movement of Afghanistan); he also had an impressive military background. A former Marshal in the Afghan army, Dostum was a crucial ally to America in the first ground campaign against the Taliban immediately after 9/11, which was later made into a movie called *12 Strong*. My job was to escort the Japanese ambassador to his residence, and Dostum was always extremely charming. A stout, ruddy-faced man, he was regarded as a celebrity as much as a politician, and it was obvious that his foot soldiers worshipped him. He was a real larger-than-life character and a key player in much of Afghanistan's rich history. One quote about the Taliban attributed to him was, 'We will not submit to a government where there is no whisky and no music.' He certainly would have got my vote, if I was an Afghan.

A lot of our work was also intelligence gathering. Whenever diplomats wanted to visit a new location, we would first have to do an in-depth recce report, providing a detailed building layout, photographs from all angles, ten-figure grid references, instructions on how to get there and alternative routes in the event of an emergency. If a location did not meet the minimum safety standards, we wrote that into the report. This could be for a number of reasons, such as being too far away from the Green Zone, or simply that it was too high-profile and had been hit too many times before – something that had been ignored in the case of the Serena Hotel.

All recce reports were stored on a database, and revised and updated as information changed. However, in such a volatile climate, offices and residences were regularly relocated without anyone telling us. On one occasion I took a senior diplomat to a large house in an affluent area we had visited many times before, only to find it padlocked and the gates overgrown with bushes.

My first eight-week stint went in a blur, and I was due some much-needed leave. However, I couldn't take time off until the guys who were currently on leave returned to the embassy. While understandable, it was also annoying and meant I almost always missed the early UK flights home. This usually resulted in nine hours hanging around at the Dubai departure gate, and after eight weeks in Afghanistan, believe me, you are more than ready for a drink. Despite that, I never got drunk there. The old me from Iraq had finally grown up. Getting home in a drunken mess when my little girl had waited eight weeks to see me was not an option. At the most, I had a couple of drinks to alleviate airport boredom, but that was enough.

The plane typically left at about 2 a.m. and I would sleep the entire flight back, ensuring I was fresh for a full day with the family. This was necessary, as Lexi was four and a real bundle of energy. She looked uncannily like Nicky had as a kid, and was also tall and super-bright like her mother.

I was on leave, relaxing as the British winter started to bite, when Nicky dropped the bombshell that she was pregnant again. I was chuffed to bits, as we wanted a family of four. Well, I did. Nicky wanted six!

She started to nest once again, buying new bedding and curtains, and transforming our spare room into a nursery. With Lexi, we had chosen not to know her gender beforehand, but now that I was in Kabul so often, we decided to be a little more organised. I wasn't too fussed either way: if it was a girl, we would already be well prepared, having had Lexi, while a boy would mean we had one of each. A boy would also be the only male child in the family, as both my sister and Nicky's sister had girls.

I was back in Afghanistan on the day of Nicky's ultrasound scan and had a few missions to complete for the embassy. Luckily, they were scheduled early and, as Kabul is four-and-a-half hours ahead of the UK, I could be back in time to take her call. Predictably, the morning dragged and as I waited for one diplomat to end a tedious meeting, my mind started wandering. Would we have a boy or a girl? What kind of parents would we be? Whatever my doubts, I knew that either he or she would be loved unconditionally and brought up with a caring family support network, just as Lexi had been. I knew that with absolute certainty. Over the years I had seen how Nicky put everything into being a mum. She had the perfect life balance when it came to parenting.

Eventually, the diplomat finished his business, and I went back to my room to dump my gear. It was cold and I flicked on the kettle for a hit of caffeine, waiting impatiently for my phone to buzz. The screen flashed half an hour later. I picked it up and didn't even have time to say, 'Hello,' as Nicky gushed, 'We're having a beautiful bouncing baby boy!'

I was ecstatic. One of each was ideal and we would finally have a boy in the family. I did have a sneaking suspicion early on that it was going to be a boy, as Nicky had been carrying this child very differently to Lexi. There was a lot of wriggling and kicking going on.

To me that was unmistakable. It was the typical sign of a restless Collins male.

16

The Fighting Season

Winter was usually quiet in Afghanistan, as it's too cold to fight. Combat in sub-zero temperatures is even more miserable than usual, and so the Taliban would knock what we called the 'fighting season' on the head and go home.

Kabul, nestling in a deep basin and surrounded by the towering white-capped mountains of the Hindu Kush, is particularly hard hit by bad weather that turns the city into an ice-bath, with snow sometimes knee-deep. A lot of diplomatic missions have to be called off, as the freezing slush on the main roads causes gridlock, even when vehicles have ice chains. Winter didn't stop all attacks, however. Just before Christmas 2014, a suicide bomber wearing an explosive vest detonated himself at a French-run school, killing several people. Ironically or not, he blew himself up during the performance of a play called *Heart Beat*, which condemned suicide attacks. The Taliban claimed responsibility.

It went quiet for a bit after that. But as the snow on the mountains melted, the Taliban regrouped, kicking off the 2015 fighting season with a major attack on 13 May.

This time the target was a concert, mainly attended by the international community, at the Park Palace guesthouse.

Shortly before the music started, three gunmen stormed the guesthouse, slaughtering fourteen people, wounding seven and seizing fifty-four others. Among the dead were a Briton who worked for the British Council and the highly respected American economic development specialist Paula Kantor. The hostages were eventually rescued by Norwegian and Afghan special forces, killing the Taliban militants in the process.

Despite the fighting season being 'officially' under way, I had a month off so I could be with Nicky for the birth of our baby. Due to the complications she had with Lexi, this time it would be a baby by appointment and she was booked in for delivery by C-section.

Thankfully, it went smoothly and on 10 June 2015 out popped Jaxson, the first baby boy of the next Collins generation. I got to cut his umbilical cord and hold him while doctors stitched Nicky up. He had the same eyes as me and I was instantly in love, as I had been with Lexi and her cute button nose. They moved Nicky into the ward and I spent the day with her, making tea or whatever else she wanted. She was a lot better this time, laughing and joking with us.

Lexi came to the hospital later that day to meet the baby. She peeked over the little cot as Jax was having a cry. I asked her what she thought of her brand-new brother.

'He is small and very cry-ish,' she replied.

Our family was complete. I had a solid month off and it was fantastic to be home with just the four of us. Jax was quite a good sleeper and would suck on his fingers to pacify himself, and we would take turns in getting to him if he cried. However, Nicky always seemed to hear him long before I did. Mother's instinct, I guess.

That brilliant month with my expanding family went way too quickly. Before I knew it, I was back in Kabul. As I checked into the embassy, the lads told me that the Afghan parliament had been attacked while I was away. The Taliban detonated a car bomb outside the building and then stormed the complex with RPGs and assault rifles, killing an Afghan woman and child, and injuring forty civilians. It hadn't even made the news back home, but stuff like that seldom did anymore. It was a case of 'same shit, different day'.

Soon afterwards, the Japanese ambassador wanted to travel to Bamyan in central Afghanistan, which was a fair distance away, so he would have to fly in on a UN helicopter. As this was far from the relative safety of the embassy, half of the team – six men – would accompany him, while the rest remained behind to run other diplomats' missions. Being fairly new, I was left behind, but I was more than happy with that, as I hated my gym sessions being disrupted. I didn't have much time for that, though, as running on fifty per cent manpower meant we were critically understaffed, rushing from one job immediately to another.

Fortunately, nothing serious happened in those few days – but that didn't last. A few weeks later the Taliban staged three co-ordinated attacks, causing maximum carnage. The first was a truck bomb at 1 a.m. near the Japanese Embassy that woke us up big time. We were all on edge, expecting more mayhem, and sure enough, later that day, a suicide bomber blew himself up near the Kabul police academy, killing twenty young cadets.

The third attack happened at Camp Integrity, which at the time housed a US special forces unit, killing one Green Beret.

We held daily intelligence briefs and would usually have the story of each attack pieced together within a few hours. Boards would then be marked with red pins noting when and where the raids had taken place, so at any given time we knew more or less what was happening on the ground.

This was invaluable intel, and one evening I was tasked to do a recce on the Kabul Star Hotel to establish if it was secure enough to house a group of construction advisors from Japan. It seemed that the embassy did not want them to live on site with the diplomats, so I needed to check if the hotel was a good enough venue from a security perspective.

From the moment I arrived, I knew it wasn't. For a start, it was not easily accessible, so any rescue mission from the Green Zone would take our team too long to get there. It was also on a rise, which exposed it to mortar and rocket fire, while the hotel security itself was not up to scratch. Even worse, I heard that the hotel held parties on Thursday nights that attracted foreigners from all over the city. With the Taliban now smashing hotels and guesthouses every week, this was a massive red flag. My recce report reflected all this, and after a lot of deliberation, as well as some internal politics, my advice was taken. The construction workers were allowed to stay on the embassy grounds, but this also ramped up the number of people we had to protect, so extra static guards were needed.

Towards the end of 2015, Kabul suffered one of the biggest attacks of the year. This time it was right on our doorstep. I had just got into bed and heard a good few rounds of gunfire go off, which didn't bother me, as that was basically background music in Kabul. But then I heard an explosion – definitely an RPG – which meant a full-on assault was under way.

We didn't know it at the time, but a residence housing diplomats from the Spanish Embassy just a hundred yards from us was under siege. Taliban fighters had used a three-storey building under construction as a firing platform, hitting the residence with small-arms fire, PKM machine guns and RPGs.

As all the action was just down the road, our team got geared up and some rushed to the ops room to get a steer on what was being hit, while the rest went onto the roof. Tracer fire lit up the sky right next to us, and the sonic 'cracks' of bullets whistling past our heads was way too close for comfort. Our diplomats were safely locked in their rooms, but we decided that if the situation deteriorated, which seemed likely, we would move them into the embassy's fortified bunker. If everything went south from there, we would have to extract them by vehicle. Extractions are a last resort: extremely messy and they usually have a bad ending involving corpses. In fact, a friend of mine nearly lost his life in one and actually phoned while bullets were flying, asking me to say goodbye to his wife if he didn't make it. I was in the UK and took the call in an Asda supermarket. As I hung up, to my astonishment, I saw his wife coming down the aisle straight for me! What could be worse timing than that? Somehow I managed to keep a lid on it and say that I was sure her husband was fine. Fortunately, he did make it.

The Spanish residence attack lasted almost ten hours. Nine people died, two of whom were Spanish military policemen. I had been saying for some time that hospitality establishments needed to be permanently out of bounds for diplomats, and once again voiced my concerns to the bosses. It wasn't rocket science, as these were easy targets. The Taliban picked them for that very reason.

Thankfully, after the attack on the Spanish embassy, management agreed with me, and all hotels and guesthouses were struck from our duty list.

Shortly afterwards I was appointed deputy team leader. It was not a role I was after, but by then I was one of the longest serving members. The other guy who could have taken the position did not want it, so reluctantly I accepted the gig, which also meant replacing the team leader when he took leave. Yes, it was more money, but for thirty days I would have eleven men to look after and to organise.

The one characteristic my team had in common was that they were all alpha males: strong, dominant personalities who would argue the toss about anything they did not agree with. This was not necessarily bad. I had figured out early on that it was often better to share a problem and then thrash out the best solution, so I ran a fairly open team. If I had to do something I didn't agree with, I knew sure as hell the team wouldn't either. However, such situations are trickier when you're in charge, as you're the middleman between the team and management, and taking shit from both sides. In my younger days I had been very much a 'people person' – the life and soul of a party – and now, having been appointed a leader, I had to resume that role. I was the go-to person, the problem-solver making sure all was well with my guys. Luckily for me, the team was solid, fairly well-balanced, and everyone was a true professional. On the odd occasion, I had to crack the whip and keep egos in check but, all in all, the team ran itself.

The next fighting season, 2016, kicked off with a vengeance, resulting in one of the biggest and deadliest single mass killings to date. An Afghan government team responsible for escorting VIPs was hit by the Taliban, killing ninety-six security people and injuring more than 300. The entire nation reeled with shock. The endemic violence sweeping the country had just been ratcheted up multiple notches, if that was possible

One thing was certain. This was going to be a bloody year.

17

The Green Zone

The sprawling Green Zone we lived in was home to a lot of different nationalities, but the common denominator was that the static security guards were nearly all Gurkhas.

These British-trained former soldiers worked in shifts, manning the gates, checkpoints and towers around the clock. In total, the Japanese Embassy employed sixty-two Gurkhas, who also guarded the British and Canadian Embassies.

Like us, they were prime Taliban targets, and in June 2016 – not long after the massacre of the Afghan VIP convoy guards – a suicide bomber killed thirteen Gurkhas about to work a shift at the Canadian Embassy. It happened while travelling from their camp into the Green Zone, when their minibus got bogged down in a crowded market area. From what we gathered, it was an opportunistic attack, as the bomber saw the Gurkhas moving slowly in a soft skin (not armoured) minibus, and simply walked up and thumbed his explosive vest detonator. This was emotionally close to home for us, as we all liked these tough Nepalese fighting men.

Another key Taliban target was the American University of Afghanistan – a location that our diplomats regularly visited. I hated going there, as it was an hour's drive from Kabul's Green Zone on a dangerous route and the Taliban were not big fans of academia.

Consequently, it was high on our watch list and we tried to limit any campus visits to thirty minutes at the most. Unfortunately, most of the trips involved ceremonies where the ambassador himself was the VIP, and these functions had a habit of dragging on forever. Academics tend to enjoy the sound of their own voices and in one ceremony I went to, they droned on for three hours. This was a massive red flag security-wise, so I voiced strenuous concerns to the ops room. Fortunately, they listened and, from then on, we were instructed to pull the plug on any campus event lasting more than half an hour, without exception.

It was a good thing we did, as a week or so later, on 24 August, the university came under a multi-pronged attack, triggered by a car bomb and followed up by gunmen spraying the campus with automatic weapons. Thirteen people died and thirty others were injured. My thirty minutes' limit stipulation for university visits was then reduced to zero.

In fact, August 2016 was one hell of a month for attacks – possibly the worst so far for civilians caught up in the bloodbath. As a result, embassy staff became spooked and cut down on official visits. I understood their apprehension. Most of the Japanese diplomats had never been to Kabul before, and to be catapulted from some sedate consulate into Afghanistan's seething cauldron of violence must have been terrifying in the extreme.

In individual bodyguard work, good client relationships are vital. It has to be a finely tuned balance of a client knowing what to expect in the event of an attack, while understanding why we need him or her to do exactly what we say, no matter how strange it seems. We lived in two different worlds, and just as I would not be so great at diplomacy stuff, they needed to trust us implicitly in negotiating rough neighbourhoods. Whenever new clients arrived, we would show them all the kit and equipment, including armoured cars and life-saving medical bags, then do a simulated vehicle attack. No live ammo was used, but otherwise it was as close to reality as possible, involving high speeds, loud shouting and even the client being manhandled a little by his or her personal protection officer. This left them under no illusions as to how dangerous it could be, and if everything went kinetic, they knew exactly what to expect from the team.

With the escalating violence, the embassy also started implementing long-overdue security upgrades. We had been asking for some time for extra bomb-resistant T-walls where locals parked their cars, as well as more rooftop firing positions.

Once again, this was fortuitous, as the vicious 2016 killing season closed as violently as it had begun. This time the high-profile target was the Ministry of Defence – and the Taliban didn't just hit the building once, but twice in a row. Fifty-eight people died, and having the sheer audacity to attack the nerve centre of the Afghan military was about as brazen as one could get. Only the presidential palace would have been a bigger prize.

Even worse for us, the MOD was right on our embassy's doorstep. The bad guys were definitely getting too close for comfort.

We also noticed that the locals were growing more fearful, and our Afghan staff and drivers told me it was no longer safe to go out at night. They just went home and locked themselves up in their homes. The markets, the social hub of Kabul life, were increasingly becoming no-go zones.

However, if we thought that the violence of 2016 was bad, 2017 was even bloodier.

It kicked off on 8 March, when a suicide bomber from the Haqqani Network blew open the back gate of the Daud Khan military hospital opposite the Green Zone, clearing a path for five gunmen disguised as medics to enter. They rampaged through the building, executing patients and staff at point-blank range, throwing grenades into wards and even stabbing people as they lay in their beds. This included women and children. The intel I got indicated it was an inside job, and the main target had been the relative of a top-ranking army officer receiving treatment. But the reality was they slaughtered anything that moved and forty-nine people lost their lives in that horrific attack. The rescue mission took several hours, with Afghan special forces landing on the roof in helicopters and clearing the gunmen out. Most of it we watched on the news, but we could hear it outside the embassy doors, where helos were flying, and gunshots and explosions blasted long into the evening.

The secretive Haqqani Network is aligned to the Taliban and has been operating since the seventies, when it first fought the Soviets with funding, ironically, from the United States. That's how screwed up politics in that region is.

One morning, two months later, I was up early and about to go for a run before the day became unbearably hot. The rest of the team was asleep, and I crept about in the darkness, looking for a pair of shorts and a T-shirt. I then went to the kitchen to grab a bowl of cereal and a decent coffee before doing some laps around the embassy ground.

I sat on a kitchen chair and looked down, as I stuck my spoon into the cereal bowl. It was a good thing I did.

KA-BOOM!

The loudest explosion I'd ever heard blasted my eardrums. The double window in front of me blew clean off the wall, frame and all, into my face and flung me backwards. Blood ran down my nose and onto the floor. Luckily, the glass had not shattered, and the fact that I had been looking down the split-second the bomb went off meant that the hardest part of my skull took the brunt of the flying debris, shielding my nose and eyes.

At first I thought the Taliban had smuggled a bomb into the building so I ran to the kitchen entrance. Shocked, I saw that every window was shattered and all doors blown off their hinges. But where had the bomb gone off? I couldn't see any blast site, which was bizarre, as an explosion that big would have left a gaping hole.

Then I heard something clattering on the roof, sounding like rain. Except this wasn't water – it was 'raining' thousands of metal fragments. I instinctively ducked. The last thing I needed was a shard of steel smashing into my head. That would hurt.

Suddenly, it all became clear. Thick black smoke spewing from a crowded intersection in front of the German Embassy right next to us told the story. A suicide bomber had detonated a sewage truck crammed with an estimated 3,300 pounds of explosives, deliberately timing it for peak-hour traffic to cause maximum carnage. The raining metal was the remains of the truck.

Every siren in the Green Zone blared. There were alarms I had never even heard before going haywire. Then there was gunfire – loud, incessant, staccato – echoing down the roads. The Green Zone, the most heavily guarded area in the world, had been well and truly breached.

My operational kit was in the bunker, so I high-tailed it across the hundred-yard compound to the side access door, yanked it open and ran down the stairs. The guys who had been asleep when I had left to go for a run were now wide-eyed and fully kitted out with body armour and ballistic helmets. It must have been the mother of all wake-up calls.

Jay, our team medic, looked up as I entered. 'Fucking hell, H. That was massive.'

He then saw I was bleeding. 'You good?'

'Yes. Nothing serious.'

We ran up to the ops room, dodging lights dangling from the destroyed ceiling panels, as I explained how the kitchen window had been blasted into my face. All radios were active so at least we had comms with each other. I remained in the ops room with half the team ready for a secondary attack – which is what usually happened – while the rest escorted embassy staff into the bunkers.

We then surveyed the damage. Bodies were strewn everywhere around the blast crater, which was a staggering fifteen-feet wide and thirteen deep. The front entrance to the German Embassy had been completely destroyed … literally blown to smithereens. It was totally open for a follow-up attack, and security staff were feverishly moving all diplomats and staff to the Resolute Support Headquarters, NATO's main base in Kabul.

Once the Japanese ambassador and diplomats in our care were securely locked in the embassy bunker, we began a sweep and clearance of the building. The windows at the entrance were destroyed and shards of glass scattered in a thousand jagged pieces. Just like the ops room, much of the ceiling was ruined, with lights hanging from wires. Inside the offices it was the same story. The place looked like ... well, a bomb had gone off. I noticed that an office worker had been injured at some point and there was a trail of blood leading out into the corridor. The fire alarms were still wailing like banshees.

It was a scene from hell. In fact, worse. The final toll was 150 dead, almost all civilians, and 413 people injured.

What worried me most was that this would soon be on global TV news and Nicky would see it. The attack happened at 8.25 a.m. local time, so just before 4 a.m. in the UK. She would still be asleep, but not for much longer. Especially with a two-year-old around.

Nicky knew that the German Embassy was right next door to where I worked, so I sent her a brief message to say I was okay and there was nothing to worry about. I would tell her about the window blowing up in my face some time later. However, that backfired, as I had forgotten all about it when I video-called her that afternoon. She took one look at the cut on my head and exclaimed, 'Oh my God! What happened?'

For the remainder of that day and well into the evening, all embassy staff hunkered down in the bunker, expecting further follow-up attacks at any minute. Fortunately, none came.

A couple of months later, on 31 July, the Embassy of Iraq was attacked. An ISIS suicide bomber blew up the gate and then three gunmen opened fire, killing two embassy workers. The trend seemed to be targeting embassies – something we were acutely aware of. We made sure we had strategies in place to repel any attack, but as the old military saying goes, 'No plan survives contact with the enemy.'

So what would happen if things really went south?

We regularly discussed this among ourselves as a team. How would we carry out a hot extraction plan? Unlike most other embassies, we didn't have a helicopter landing site, so an airlift was difficult, if not impossible. Consequently, the only way out in an Alamo-type situation was by car. But as mentioned, vehicle extractions are usually messy, with bad endings. In any event, with twelve guys in our team, only six cars could attempt an escape at any given time, as we would need each vehicle to have a driver and a commander. We had more than eighty people at the embassy, including the Gurkhas and operations staff, so that was out of the question.

But even if we had enough contractors and vehicles, we would need to shoot our way out to the airport, optimistically assuming it would be open. Otherwise, we would be running the gauntlet to one of landlocked Afghanistan's borders.

But which one? None of us had visas to enter any neighbouring country anyway.

I wasn't an excessive worrier. But this was enough to give anyone nightmares.

18

Tying the Knot

Snow started falling on the Hindu Kush mountains and the Taliban went home. The 2017 fighting season had slowed down and Christmas arrived within the blink of an eye.

I would, once again, be spending this festive season with the lads at the embassy. We all brought in a load of treats and snacks hoarded from when we had last been home, as well as a few bottles of grape juice – being on permanent standby meant strictly no alcohol. Even though we were away from loved ones, Christmas Eve at the embassy was always a proper good party. We had to make the most of what was really a shitty situation.

Thanks to modern technology, I watched my kids open their presents on a video call the next day. Even though not physically there, it gave me intense pleasure to see those happy little faces. I know it's overkill these days, but in a way I enjoy spoiling my kids. I'm not a material person and nor is Nicky, so it goes on the kids.

Another expense Nicky and I had put off for a long time was getting married. Neither of us was too fussed about it, but we wanted the family to share the same surname. It didn't feel right that Lexi and Jax were Collins, and Nicky wasn't. Also, after putting up with me for all these years, I felt that it was time to entitle Nicky legally to half of everything I owned.

We began planning for a small, intimate wedding with minimum fuss. All I had to do was start organising my side of it, from halfway around the world. No worries. Nicky went dress shopping with her mum, and her parents reserved a lovely venue and paid for the guests to stay over. We booked a band and sorted out the menu.

I'm sure many will agree that one of the great things about weddings are the hen and stag parties beforehand – a chance for friends and families to let off some steam and party together. Ours was no exception. Nicky and her friends went to Dublin for a weekend with all the usual shenanigans, such as real-life nude drawings, male strippers and cocktail-making. There were about fifteen of them and from the videos I saw, it looked like a bloody good bash.

I decided to do something different, ticking off a top item on my bucket list and riding my Ducati 848 up to the Nürburgring racetrack. I couldn't think of anything better than three days on that iconic circuit with a helmet on and no one talking to me as the miles flew past in a blur. Bliss! The journey itself involved a fair stint in the saddle, first up to Hull to catch the Channel ferry to Holland, then riding all the way into Germany. Accommodation was an easy choice, as I had heard of the famous Sliders Guesthouse, a hangout for Nürburgring nuts. Epic stories of fun and revelry at Sliders are cemented in biker folklore and the English owner Brendan Keirle was an absolute legend.

Nicknamed the 'Red Baron', in the biking world Brendan was the equivalent of *Top Gear*'s Stig. He booked me in for four nights and updated me regularly with all the good places to go for food or beers, and which car or bike museums were worth a look. He also offered to show me the ropes once I got there, as the track is massive and somewhat overwhelming for first-timers.

Sadly, that never happened. Just before I arrived, Brendan was killed on the Nürburgring track he loved so much. His car had developed an oil leak, and he pulled over to warn other riders and drivers. As he got out, one vehicle hit the slick and lost control, ploughing into Brendan. He died instantly. It was a particularly tragic and cruel twist of fate, as he was set to retire a few weeks after my visit.

I was shocked at the news. Death seemed to follow me wherever I went, even outside war zones. It was a blunt reminder of how precious life is.

Riding from Rotterdam to Nürburg with Google maps on my phone and my headphones blaring Whitesnake – always great music for the open road – I was at peace with the world. Being on my bike was one of the few times I got lost inside my own head, thinking about everything that had gone on in my life and wondering why I still sought out dangerous pursuits. Brendan's death made me question my choice of running the Ring, especially just before my wedding, but ultimately it was the same as my work and sometimes people died. We all knew the risks and the song remained the same. Why would this be any different?

On the first day it hammered down with rain, making the Ring particularly dangerous. I went to the track with a lad also staying at Sliders, who took me for a few laps as a car passenger. After that, I strolled around the parking lot and got talking to a Russian dude. He had a Nissan GT-R and offered me a few laps, again as a passenger. It was awesome and although the track was wet and slippery, it was still pretty fast.

The weather cleared enough on the second day for me to attempt some tentative laps on my own. While buying a ticket, I bumped into some guys running Honda Fireblades and asked if I could use them as my marker. They agreed, as they had already done a few laps, so they had some experience. We spent the rest of that day riding the track, stopping for lunch at the Devil's Diner, and then wandering around checking out all the cool bikes and cars. Bennet, my best mate from childhood, was also there, on an old Triumph, and we loved looking at the huge variety of magnificent machines.

On my last day it rained yet again and I sat trackside, patiently waiting for the skies to clear. I was desperate to put in a lap before riding back home and eventually got my wish, doing a final run. This time, being a little bit more sure of the circuit, I opened up the Ducati and caught some riders ahead. As I followed them, one suddenly came hurtling off his bike and smashed into the barrier. His mates immediately stopped to help, and I sped ahead. Soon afterwards, the track was closed and an air ambulance flew the injured biker to hospital. Pulling into the Devil's Diner, I flipped up my visor and said to Bennet, 'Now's the time to head off.'

The bucket-list itch of riding Nürburgring had finally been scratched.

At last our wedding day arrived. I had been with Nicky since 2005 and we had been through a lot together. The time was most definitely right to tie the knot. In fact, it was perfect. The day itself was one of the best of my life and passed almost in a daze, as our feet barely touched the ground. Nicky looked absolutely stunning. She had put a lot of thought into her dress and it was very 'her': beautiful and classy. The ceremony was small, and we exchanged vows in front of a select group of family and friends. Lexi and her cousins were the bridesmaids, and Jax was a pageboy.

We spent our honeymoon in Edinburgh, one of our favourite cities. That year was a rarity for me; celebrating Christmas and New Year at home, and having just got married made it even more special.

It ended all too soon. It was time to go back to the war zone.

And reality.

19

The Good Doctor

It was January and bitterly cold when I arrived in Kabul for the first time as a married man. But apart from workers making a racket while fixing the bomb-wrecked sections of the embassy, not a great deal was happening. Hopefully the quiet season would continue.

Some hope. On 20 June 2018, four men armed with AK-47s and RPGs stormed the Intercontinental Hotel, killing forty people. I watched it on the local news channel, and it was harrowing in the extreme to see petrified people tying bedsheets together and trying to climb out of top-floor windows as fires raged. It took the security forces twelve hours to rescue 153 hostages and kill the militants.

Among the dead were two Americans and a German. By then all hotels were out of bounds for our embassy staff, but for some reason, other foreigners still frequented establishments specifically targeted by the Taliban, ISIS or the Haqqani Network. Absolute madness.

A few months later, in November, the private contractor community suffered a terrible personal loss. I had just arrived back in the UK, and as it was approaching our first wedding anniversary, Nicky and I decided to go to

Edinburgh. We had honeymooned there, but I had come down with 'man flu' and so we hadn't enjoyed it as much as we should have.

On the way to the station, my phone kept vibrating. I ignored it, as it was put away in my daypack, and besides, I was on holiday. Eventually, Nicky unzipped my bag.

'Here,' she said, giving me the phone.

Instantly, I noticed five messages, all about the same thing – a compound called Camp Anjuman that provided secure accommodation outside Kabul had been attacked. That certainly got my attention, as I knew a few lads at the camp, and one in particular – Luke Griffin – was a good friend.

I banged off a text to my lads, asking if anyone had news of Luke. A reply came back almost instantly: 'Have messaged Luke but he has not read it.'

Luke was one of my favourite people. Ex-military and a skilled operator, he would be in the thick of the fighting but I was sure that when the chaos calmed down, he would get back to us and let everyone know he was okay.

Nicky and I arrived in Edinburgh, checked into our hotel and turned in early to be fresh for the next day's sightseeing trips.

I woke before dawn broke and for some reason reached for my phone. The screen lit up in the darkness and I squinted to read the top message. My blood ran cold.

'Mate, Luke was killed last night in the attack. I'm not sure at this point how many people know, but I thought you guys should know. Gutted doesn't even come close.'

I lay back in bed, drained, staring at the ceiling. A feeling of total emptiness swept over me. Luke was a husband and a father – in fact, his son was the same age as Jax. He was such a high-spirited character: a genuine, straight-up bloke who was always there for his friends. I first met him in 2015 when we were working for different security companies and ended up at the same function. It was a long night, and I mentioned I was so hungry I could eat a horse. Luke had contacts everywhere and within minutes had organised a memorable feast; we got on like a house on fire, discussing former military units and various gigs in the badlands.

My thoughts then turned to Luke's best friend, Alek, who was on our embassy team. They had been on previous contracts together and were pretty much inseparable – a pair of lovable rogues who could find mischief anywhere. Alek was among the strongest guys at the embassy, never without a roll-up cigarette in one hand and metal mug of tea in the other. Extremely sharp-witted, you had to be up early to get the better of him.

I sent a message across to Alek, knowing he would be distraught.

A few hours later, updates of the Anjuman firefight started filtering through. After a car bomb had ruptured the outer perimeter, a group of four heavily armed Taliban militants firing AK-47s and tossing grenades stormed the camp. About to be overwhelmed, Luke directed his clients to safety and then engaged the gunmen. Despite being outnumbered, he managed to slow down the attack, buying precious time and saving lives, until he was fatally wounded. He had been an absolute hero and a warrior to the end.

All of the lads at the embassy who knew Luke were distraught. It was a truly dark day.

But for our embassy clients, the most traumatic incident of all occurred at the end of 2019, when one of

Afghanistan's most revered doctors, Tetsu Nakamura, was gunned down.

A Japanese national, Dr Nakamura was based in Jalalabad and his company, Peace Japan Medical Service, operated clinics for leprosy patients and refugees, as well as supplying clean water for scores of villages afflicted by cholera. He had been made an honorary citizen of Afghanistan, and in 2003 won the Ramon Magsaysay Award, widely regarded as the Asian equivalent of a Nobel Prize. The locals called him 'Uncle', and he was arguably the most celebrated foreigner in the country.

On the morning of 4 December 2019, the seventy-three-year-old doctor was being driven to work when five men stepped into the road and opened fire, instantly killing his security guards. They then shot Dr Nakamura in the chest. As he lay bleeding in the dirt but still alive, the killers said, 'It's done, let's go.'

By then the Japanese Embassy in Kabul was on full alert, instructing us to speed to Jalalabad for a casualty situation. Kabul to Jalalabad – or J-Bad as we called it – was a good three-hour drive and I would be heading up the emergency recovery team.

Tragically, Dr Nakamura, who had been taken to a medical facility at Bagram Airport, died from loss of blood as we were about to leave. The body would be flown to Kabul.

I led a two-vehicle convoy escorting the embassy's head medical officer and the ambassador, a close friend of Dr Nakamura, to Kabul's airport to fetch the body. US soldiers waved us through the Taipan Gate to Ramp 5, but for some reason we were held up there for more than two hours while waiting for instructions from Tokyo.

Eventually, we were escorted to a sectioned-off ward in the US military hospital, where the chief medic pulled back the curtains. In front of us lay Dr Nakamura. Distressed, the Ambassador took off his glasses and held his hand to his eyes. On the bed before him was a good man – a great man – who had been killed in cold blood.

'Give us a minute please, guys,' I said, ushering the Americans out of the ward. Ben, the ambassador's private protection officer, and I stood back, giving the tearful man some space as he formally identified his friend. Despite never having met Dr Nakamura, I had a lump in my throat. Even in death, he seemed to have that effect on people.

I left Ben with the grieving ambassador as I discussed with the Americans what procedures were needed to release the body. They said they had a fully operational morgue in the hospital that the embassy was welcome to use. After we had completed the ID, I explained to the ambassador that the body could remain where it was and we would have access whenever we needed.

'No,' he said. 'We will take him to the Kabul military hospital.'

This was the same hospital that had previously been attacked by the Haqqani Network, killing forty-nine people. Unsurprisingly, I was not mad keen on that idea, but there was little I could do. It was an incredibly delicate situation and at that point I didn't fancy explaining to the distraught ambassador the serious security risks involved in moving the body. I rang operations and updated them that we would be doing an additional stop, or 'bounce move', and sent across the ten-figure grid reference for the Kabul military hospital. They were dumbfounded. The considerable risks in the bounce move were totally unnecessary. But like I said, sometimes you just have to go with the flow.

We loaded the doctor's body into an ambulance and I had a quick briefing with my team, outlining which route we would take and what to do if we were attacked, and telling them to cover their faces, as there would probably be media at the hospital.

The journey took ten minutes. As we arrived, I noticed a group of people outside, smoking. I assumed they were waiting to carry the body from the ambulance.

'Awesome,' I whispered to no one in particular. 'No press.'

I could not have been more wrong. The group I had thought were hospital orderlies were actually journalists. Before I knew it, cameras were pointing everywhere, as reporters and photographers jostled each other to get a snap of their beloved 'Uncle's' body. I shoved them out of the way and shot one guy a menacing look, which made him pay attention somewhat sharpish. He said something in Dari and the cameras were lowered.

I nodded at him. '*Tashakur.*' Thank you.

Although grateful the press had complied, inside I was seething. They were at best a bloody nuisance; at worst, a serious threat to our safety in exposing our identities. We placed the body in a room and stepped aside while the ambassador and a high-ranking military official had a discussion. It was in Dari, but from what I could gather, the body would definitely be kept there. I groaned inwardly. Not only was that a major security problem for us, but also a practical one for the simple reason that the hospital had no freezers. I could only see an air-con unit, so asked the military officer if he had the means to keep the body cold.

'Of course.'

'With dry ice?'

'Yes.'

I was not convinced, but it was the embassy's call. As we were about to leave, the military officer came charging over.

'Mr H, do you have any dry ice?'

I stared at him. I almost wanted to say, 'Yes, mate, what colour?' but bit my tongue instead.

'No, but you're going to need to get some quickly,' I said, jumping into my vehicle and reversing out.

That wasn't the end of the saga. I guess that guy never did find the dry ice he claimed to have, as the next morning I discovered that Dr Nakamura's body had been moved yet again to another location. The only problem was that no one knew exactly where.

Speaking to the local drivers, we discovered there was a clinic in the south of the city that dealt with embalming and cold storage. Perhaps the good doctor had been moved there. We had no other leads, so acting on this sketchy info, we mapped out a route to roughly where we thought the clinic was. Effectively, we were going in blind, which is possibly the worst thing to do in terror-gripped Kabul. But we had no option.

Luckily, we found the clinic, and yes it did have cold storage. And yes – they had the body!

Breathing huge sighs of relief, I flashed our diplomatic licences and was let in. It was a fairly run-down affair, which was not unusual in Afghanistan, but then the strangest thing happened. One of the team guarding the vehicles radioed me.

'H, you have a reporter inbound.'

I stopped dead in my tracks.

'For fuck's sake!' I muttered, quickly instructing my guys to cover their faces, as we braced for the usual Afghan media circus of jostling journos and camera flashes.

Not this time. In fact, the reporter was not an Afghan, but a well-known Japanese journalist. I was dumbfounded – we didn't even know she was in the country; how had she managed to track down this obscure embalming clinic, while we had had to use guesswork? I still don't know the answer.

I quizzed her credentials and it turned out she had been a good friend of Dr Nakamura's, so I allowed her in. I am not mentioning her name in case she is still in Afghanistan.

Now that I knew where the body was, I drove back to the embassy to fetch the ambassador. When we returned, the journalist was still there and the ambassador was as surprised as I had been. However, he seemed to know her, and all I could guess was that the authorities were keeping her presence in the country a secret.

From there, events unfolded rapidly. The doctor was getting a state funeral – an indication of how celebrated he was – and it all had to happen the next day.

The first task was to move the doctor's body to the Kabul military hospital, where the Nakamura family, who had flown in twelve hours earlier, would say their final goodbyes privately. We moved off early to miss peak-hour traffic, getting to the clinic in less than thirty minutes. It was extremely fortunate that we did, as almost immediately we hit our first problem. The bullet that had so cruelly taken the doctor's life was still in the body. As airlines will not transport anything that sets off metal scanners, we had to get that chunk of lead out right away. This was news to the medical officer, and we had to wait patiently as the AK-47 round was extracted. All the while, the clock was running down for that afternoon's high-profile state funeral.

Eventually, an Afghan wearing surgical gear came up to me and said, 'Okay, it's out.'

I was not going to take his word for it. The funeral was going to be broadcast live and the last thing I wanted was embarrassing glitches. The bereaved family already had enough to contend with.

'Show me,' I demanded.

I followed him down to the lower floor and he waved a metal detector around the body. It didn't beep once.

'Thank you.'

He then produced some paperwork. 'Sign, please.'

At first I didn't understand what he was getting at. Then it dawned on me. He wanted me to sign for the body.

I shook my head, saying, 'Sir, I think you need to do that. I have a British passport and it would have to be someone from the country the body is flying from who signs it over.'

I was extremely concerned that this would spark a stand-off and further delay us. Fortunately, he just shrugged and scribbled his signature on the official forms.

I was still on schedule, although barely, and had to contend with the mid-morning traffic chaos in downtown Kabul. We rendezvoused with the other half of the team, who by then had secured the military hospital, did the changeover, and the family were able to pay their last respects to a man who had made the ultimate sacrifice for the people of Afghanistan.

My team returned to the embassy, stripped off our body armour and turned on the local news channel.

The funeral was aired live and I watched President Ashraf Ghani carry Dr Nakamura's coffin as one of the pall-bearers. It was an emotional ceremony. The love and respect for the man that Afghans called 'Uncle' was poignantly obvious to even the most cynical observer.

I shook my head in sorrow. To cold-bloodedly murder a man who had done so much for the poorest of the poor in an already poverty-stricken land seemed an enormous own goal.

But then, I don't think that would bother the Taliban. Not one bit.

20

A New Enemy

As 2020 dawned, a new enemy appeared on the horizon. This time without grenades, bullets and explosive vests.

Instead, it was a virus that ended up killing millions more people globally than all the Taliban, Haqqani and ISIS gunmen and suicide bombers combined.

Covid-19 had actually arrived in the final months of 2019, but its rocketing escalation only became starkly apparent in February 2020. Ripple effects soon became tsunamis, and the entire planet was effectively locked down. Everyone's world was turned upside down, but for us as security contractors in Kabul, the main problem was the cancellation of international travel. We were, for all practical purposes, prisoners in Afghanistan.

This meant I had no idea when I would see Nicky and my kids again. It was a bitter pill, but one I had to swallow, as there was no other option. So I knuckled down, remained as positive as I could be and continued to work out at the gym, eat well, keep myself fit, and generally stay on top of my game mentally and physically. Every night I watched the depressing 10 Downing Street Covid briefings and simply took life a day at a time – something I was used to in frontline situations.

In fact, it was probably a lot harder for people in the UK than it was for me. On our video calls, Nicky shed a few tears on a couple of occasions. She is an exceptionally strong person who seldom cries, but she loves her family and not being able to visit her parents living close by really got to her. It was an indication of just how much tougher the epidemic was for her, and many others like her, back home.

All in all, we ended up being locked down in Afghanistan for almost half a year, and as missions in the city were drastically reduced while the epidemic raged, it was extremely boring.

Finally, restrictions were fractionally eased and I managed to fly home in mid-July. I spoiled Nicky and the kids like crazy, as I had missed them more than words could tell. But although going home was wonderful, travelling there was pure hell, requiring countless bits of paperwork, PCR testing and even letters confirming I worked at the Embassy of Japan. The flights themselves were even worse – eight hours wearing a mask and unable to walk about or get a drink on the plane.

One thing for sure was that the Taliban were not deterred by lockdowns, let alone social distancing. Suicide bombings and crowd massacres continued unabated. In fact, the 2020 fighting season started in May with a particularly vicious attack on a Doctors Without Borders maternity clinic, where twenty-four people were killed. I shook my head in sorrow. Who would kill mothers and newborn babies? The country really had become a shitty place.

Another awful attack during the height of the pandemic was at a location we knew well: the University of Kabul. A suicide bomber drove a car onto the campus and, as it exploded, gunmen stormed the campus, killing thirty-five people. In any other country this would have been a national catastrophe. Tragically, in Afghanistan 2020, it was once more a case of same shit, different day.

It was around that time that I started seriously thinking about getting out of the security game altogether. I had been pushing my luck to the absolute limit for more than twenty years, but now I also had to take into consideration that I was a husband and a father. I had enough saved to take some time off and job-hunt, so I started putting together a few options that didn't necessarily involve being shot at. My family was the most important thing in my life. In fact, with Covid restrictions, I even got caught up in the whole home-schooling thing when on leave and set the kids up with a desk and a laptop each. They were in their element, and although I enjoyed it, I doubt I would ever make it as a teacher in the real world. It was a tough enough gig with only two pupils in my class.

I didn't know it at the time, but any career decision would soon be taken out of my hands by events way out of my control. On 3 November 2020, Donald Trump was ousted in the US presidential elections and replaced by Joe Biden. We all knew Trump wanted to withdraw from Afghanistan, but what would Biden do?

We soon discovered. On 14 April 2021 he announced to the world that there would be a 'complete withdrawal' of all American forces by 11 September 2021 – twenty years to the day that al Qaeda flew planes into New York's Twin Towers. This meant in effect that all NATO troops would also leave the country.

That was less than five months away, and I expected a lot of discussions and think-tanks at the embassy on what to do when the Americans left. At the very least, we needed a detailed blueprint for a possible evacuation and timelines for what actions needed to be taken.

Surprisingly, that didn't happen. In fact, no one at the embassy mentioned anything about it to us. This was bizarre, as even if Japan remained in the country during a possible extremely messy Taliban takeover, we still had to consider that without the Americans we would lose all our medical facilities, as well as the ability to airlift the diplomats out if everything went south. We would be completely isolated. Surely, we had important security options to consider?

Apparently not.

Another uncertainty for us was that the year before, a new security company had won the contract for the embassy, and as the ambassador and diplomats liked the work we did as a team, we were included in that contract. However, some hiccups still needed to be ironed out. For example, we had been promised brand-new M4 assault rifles, but instead got twelve AK-47s. This meant that my team was armed with weapons of two different calibres – not ideal in a combat situation, as it's vital that ammunition can be shared.

We were then invited to a tour of the new company's headquarters, located near the airport. It was a vast compound with two big high-rise buildings and several smaller accommodation sites. Local Afghans guarded the entrance to the compound, which also had a search bay for vehicles, and then another gate led to a housing area for HQ staff.

Our guide took us up onto the roofs of the buildings, and showed us the key positions and shelters, including bunkers and safe rooms. It was a good vantage point, but at the back of the compound was a building ringed with razor wire.

'What's that?' I asked.

'It's a prison,' replied the guide. 'Taliban, ISIS and Haqqani militants are in there. There's even a Westerner, caught selling alcohol.'

I frowned. Right next door to our HQ we had a prison rammed with bad guys. Maybe I was being over-sensitive, but to me that was a significant red flag.

On the plus side, the roof had decent fire points and any indirect fire, such as mortars or rockets, could be easily pinpointed. With the grand tour completed, we headed back to the embassy.

Fear and uncertainty started running rampant, as everyone could see the Americans packing up and getting ready to leave. The Afghan security forces had been trained by the US military and on paper had all the equipment needed to continue the war – but would that actually happen? Would Afghan soldiers fight, flee or desert en masse? Did they have the iron will, the raw courage, to withstand a Taliban onslaught?

That was the question everyone was asking. What was actually in store for Afghanistan?

We would soon find out. In fact, sooner than we thought. On 8 July, President Biden announced that the American withdrawal would be completed by 1 September, ten days earlier than originally scheduled.

Predictably, more chaos and confusion engulfed the land, and the Taliban went on a full-scale offensive, killing close on 3,000 civilians, as they started chalking up one victory after the next. The first district overrun was Nerkh in the east, and by 21 July – just thirteen days after Biden's announcement – the Taliban controlled more than half of the country. Zaranj in the south was the first provincial capital to fall, followed by the prized cities of Kunduz and Mazar-i-Sharif in the north. From there, the Taliban stormed triumphantly into Jalalabad, effectively surrounding Kabul.

Foreign embassies watched, alarmed, and then they started to close, one by one. First out were the Scandinavian countries, then the Dutch, Swiss and Germans, followed by the Canadians and the British. At the US embassy there was feverish activity, as Americans worked around the clock to avoid a repetition of their humiliating helter-skelter withdrawal after the fall of Saigon forty-six years previously.

The only embassy that seemed to be delaying was that of Japan. They had invested so much in the country, from educational to humanitarian projects; it was clearly a very difficult decision for Tokyo to take.

21

Last Out of the Green Zone

Finally, there was a decision. With the provinces falling like skittles and the Taliban poised to storm into Kabul, we got the order on Saturday 14 August to evacuate the embassy as quickly as possible.

We would be the last diplomatic mission to leave, apart from the Chinese, who supported the Taliban anyway. It was not a moment too soon.

In fact, considering the speed of the Taliban's advance, it might even be too late.

It would be a two-stage evacuation: first moving all diplomats and staff from the Green Zone to the security company's HQ compound as a staging area, and then to the airport. We could not waste a second, as the security situation was critical. Kabul would soon be overrun.

As we pulled up to the embassy entrance, I saw a massive pile of boxes stashed up outside the building. We already had restricted space in the vehicles, let alone what limited luggage evacuees would be allowed to take on the plane, so I asked one of the diplomats what was inside the boxes.

'Printers,' she replied.

Seriously? Computer printers? I almost exploded. I had to be careful with my words.

'Ma'am, just grab your bags and get into the car. We don't have time or room to be loading unnecessary items.'

I shoved the boxes out of the way with my foot, trying not to show my anger.

'But the Ambassador said …'

'Ma'am, with all due respect, the diplomatic mission is over. Get in the car with your passport and your bag. We're leaving now.'

The six armoured SUVs were lined outside the entrance, engines revving and doors open. It was going to be a hell of a long night, moving a total of eighty-one people. The twelve diplomats were our priority and we had to get them on a plane as soon as possible. Thankfully, the ambassador himself was out of the country on other official business, but constantly asking for updates and doing what he could to ensure the safety of his diplomats.

I walked down the SUV convoy, counting the diplomats and asking to see each one's passport. After moving them out, the plan was to return and pick up as many Gurkhas and other staff as we could squeeze in. By my calculations, we would have to do four trips. My guys would be on the final trip with ops room staff and would shut everything down.

With the count done and everyone's passports checked, I got into the sixth vehicle. The control vehicle is usually at the back of a convoy, as that's the best position to support other cars in case of an attack or breakdown, and also to dictate speed. I radioed the other vehicles.

'Six is good. Let's move.'

I looked at my watch, a G-shock that had been with me on many adventures. It was exactly 10.37 p.m. The extraction had finally begun.

Everyone in the team knew the way to the security company's HQ inside out, including the shortcuts. But even so, the safest route was always the one passing large bases or facilities, because if we were attacked, we could duck into them. We first passed the Canadian Embassy, where the usually brightly lit entrance was pitch-black. It was deserted, with six or seven discarded sets of body armour strewn on concrete planters.

'Looks like we're late to the party,' I said. Ben, the ambassador's private protection officer, who was driving, nodded quietly. If the diplomats didn't know that this mission was down to the wire, they did now.

We then passed the World Bank building, which had also been vacated. Bags and suitcases had been dumped outside, contents half-emptied, testimony to an extremely chaotic exodus. Worse, security gates usually guarded around the clock by Gurkhas and sniffer dogs had been abandoned. In fact, all of the Green Zone gates were now unmanned. I radioed our ops room, telling them that we were massively exposed, as all other facilities had been vacated.

Eventually, we reached the Red Zone border, where two local guards sitting on chairs casually waved us through. I guessed in a few hours' time they would be gone, or else they were simply waiting for us to finish evacuating the embassy so they could go on a looting spree.

We headed north-west. The route I had selected was about eight miles long, and with delays of loading and unloading, each trip would take roughly an hour. We moved as fast as we could, but I had briefed the lads to be sensible. The last thing we needed was an accident and some locals mown down, but even if that happened, we would not stop for anything. As we reached the HQ gate, I glanced at my watch. The trip had taken twenty-six minutes, which wasn't bad at all. Hopefully, we would have everyone out of the embassy before sunrise.

Despite having radioed ahead and the ops room at the embassy having warned the company that we would be arriving soon, there was no one to open the gates. The driver in the first vehicle impatiently honked the hooter. A golden rule in the Red Zones was not to 'break the seal' – open the doors – if we had clients inside, so all we could do was hoot. Eventually, HQ guards realised that twelve diplomats were in the vehicle and swung the gates open. Suffice it to say, there was quite a bit of swearing from my guys.

We got the diplomats out and left as quickly as we had arrived. Too quickly, in fact, as one of the guards over-zealously shoved the gate open and it swung back straight into my vehicle. My guys were furious and gave the guard both barrels of verbal abuse. I jumped out to calm everyone down, then saw that the bumper was totally smashed. We would have to yank it off, resulting in more time being wasted. I looked at my watch, starting to think that fate was conspiring against us. Every delay increased the danger we were in.

Eventually, we managed to wrench the mangled bumper off the vehicle and the guard continued to stare blankly after us as we sped off, still cursing.

We took the same route back to the embassy, with the convoy zigzagging across the road at one stage against a one-way traffic flow, as we re-entered the Green Zone. To my surprise, the same two Red Zone guards were still there to let us in. I gave them a thumbs-up. Then, at the deserted Canadian Embassy, I saw that one of our site-security managers and some Gurkhas had arrived to provide back-up for us. We waved at each other as the convoy sped past. He was a brave man, reminding me of Sergeant Smithy at Ramadi, as both were as mad as a box of frogs. My team would pick him and his Gurkhas up on our final trip.

We reached the embassy and I shouted, 'Five minutes, boys. Then we go.'

That gave them time to use the toilet or to make a cup of coffee, while I supervised loading the Gurkha guards. Good job I did, because some of them had massive suitcases.

'Guys, minimal kit,' I said, exasperated. 'No big bags.'

I'd already had this conversation with them, but even so, these guys looked like they were heading off for a two-week holiday to Tenerife. There was no way we could get all their kit into the tight confines of the SUV, so I simply dumped stuff, annoyed that once again we were wasting precious time.

The second run was smoother, as this time the HQ gate was open and the guard managed not to rip off the back-end of my car. We did another 'turn and burn'; at the rate we were going, we would need only another two trips. As long as everyone was at HQ before sunrise, I would consider it a huge win.

The third run was also hassle-free. Just one more to do. This time it would be the remaining members of our team, ops personnel and the security manager, and the Gurkha team at the Green Zone gate.

We had a bit of spare time while ops sanitised their office, bringing out their laptops and destroying everything else. Caffeine was required, so I pilfered a Black Rifle Coffee bag from another dude and flicked on the kettle. It was strange not seeing Fluffy, the cat, asleep on the bench outside my room as I closed the door for the last time. Earlier that day one of the lads had managed to get all the embassy cats into a cage and off to the Nowzad animal charity. It had cost a small fortune, but was worth it, as they would be rehomed in the UK, rather than starve on the streets of Kabul. Fluffy and his mates had got out before the rest of us.

As we were about to leave, we decided to take a group photo of the extraction team on the steps of the embassy. It's a pretty cool picture, taken in the middle of the chaos, and you can see some tired-looking faces – mine included.

Rob, the embassy project manager, then shut down all generators, closing the electronic gates manually. Watching him, I said, 'Last time we see this place, boys.'

I was partly sad, but mostly relieved. Afghanistan had been quite a journey – arriving in 2002 as a Royal Marine, then as a contractor in 2014, and ending seven years later with the last nerve-racking embassy extraction in a city about to be overrun. The tables had been well and truly turned; I had been part of the spearhead that chased the Taliban out, and now the roles were reversed. This would probably be my last war-zone gig, but I had always believed it would end on my terms. Instead, I was on the run.

On Sunday morning, 15 August 2021, just before the sun rose, we were the last foreigners to leave the Green Zone. The first part of the embassy extraction had gone smoothly, but this was no time for any congratulatory backslapping. The second, equally crucial, stage still lay ahead – getting everyone to the airport, alive.

Knowing that the Taliban would soon be swarming into Kabul, I was not that confident that we could pull it off. But looking at the team riding shotgun with me, I reckoned that if anyone could do it, we could.

22

The Airport Run

Gunfire crackled around the outskirts of Kabul, like Guy
Fawkes on steroids. Taliban forces were within firing range
of the city and I was getting more and more concerned
about our diplomats' safety. If we didn't get them to the
airport on time, who knew what would happen.

Despite being exhausted after the tense all-night
extraction runs, the team and I were called in for a meeting
with HQ staff. As I entered the conference room, I saw
some comfy chairs and slumped down, desperate for sleep.
A guy called Jack arrived to brief us. The main topic was
how they planned to evacuate the several hundred people
already in the compound, which was obviously important,
but all I wanted to know at that stage was exactly when
would I be taking our diplomats to the airport. My clients
were my overriding priority.

I waited for Jack to finish, then asked about the security
set-up in the compound. I pointed out that my team
was housed in building five and we had not seen anyone
guarding anything other than the gate we came through.
And even that had not been manned at the time, involving
some serious hooting to get attention.

Jack was evasive. 'Listen, mate, we're all tired and it's been a long night. We can talk about that later.'

I looked across at Rob as if to say, 'Is he for real?' Compound security was by far the most pressing issue at that moment. I was about to take it further, but instead bit my tongue and pulled out my mobile phone to send a message to Ben.

'Need to man building five with NVGs (night-vision goggles) right away.'

Ben quickly responded, 'Roger,' and we had our own guys guarding the roof of our accommodation block before Jack had finished his briefing. I never had to micromanage my team. If I asked for something, it happened – particularly with Ben, who was ex-army, strong and a good athlete, with operational experience in Afghanistan. Definitely the guy you want on your side in a fight.

I headed off to my room to crash. Hopefully, we would soon be taking the diplomats to the airport: the hairiest stage of the evacuation now that the Taliban were on the brink of entering the city. I needed to be on full alert.

A few hours later, I woke, still exhausted, and my morning was made even worse with a cup of instant coffee. I mean, who drinks that shit, unless you have to?

More gunfire erupted and I stood, tossing the instant crap into a basin. The distinctive clatter of AK-47s had significantly intensified and ominously appeared to be coming from the vicinity of the airport. If Taliban fighters were there, it meant we had lost any initiative we might have had. We urgently had to get our diplomats on a plane.

I phoned ops staff to see if they knew what was happening. Fortunately, they were on the ball. They had been calling around and had an update

'The diplomats will fly out with the Yanks at 14h00, H. An Apache gunship will escort you guys to the airport's Taipan Gate. Drop them off and head back.'

At last, we had a scheduled flight. That meant we would be leaving for the airport in a couple of hours.

In the interim, we were taken around the compound to inspect the security of every building in the event of an attack – a scenario growing more likely by the hour. The first thing that spooked me was a sort of shack that had been built on the helicopter landing site, making an emergency airlift virtually impossible. When I asked what it was, I was told it was a pizza hut. I couldn't believe my ears.

Also with us was a guy, called Rupert, who wanted to know how my team would get from the roof of building five to their offices in building six in an emergency. I asked if they had an upper gangway – a pretty basic security precaution that interlocks the top stories and allows defenders to dominate the high ground. They didn't, but Rupert had a lightbulb moment. He pointed to the gap between the two buildings.

'Do you reckon you guys could jump that?'

I stared at him in astonishment. The gap was about ten feet wide and six floors high. In a combat situation we would be weighed down with rifles, machines guns and plenty of ammo. Did this guy think we were Superman?

'Er … probably not,' I replied. 'Unless we want to plunge to our deaths.'

'What if I get some ladders? Then if we get attacked, you guys can spread them across and crawl over to building six?'

I was speechless. For a start, Bry, the team's crack machine gunner, weighed eighteen stone in his socks – and Einstein here wanted us to crawl over cheap local ladders between eighty-foot-high rooftops in the middle of a firefight.

I had heard enough. I headed off to regroup with the team, letting them know about the pizza hut and Rupert's ladder brainwave. It got a good laugh.

The volume of gunfire was steadily increasing. I still couldn't see where it was actually coming from, but it was definitely getting closer. We were all on edge, sensing that any brief window of opportunity was fast slipping away.

Eventually, Rob beckoned me over. 'H, the Apache gunship will be on station ten minutes before you move. He just needs to lock on to your vehicle.'

Happy days! I already had a fluorescent orange marker panel on the car roof and gave it a good wipe to make sure the pilot could see it. It was reassuring to have gunship cover now that the city was going loud, but I wasn't sure what the pilot's terms of engagement would be if we were attacked. We would probably be on our own.

We ushered the clients into the SUVs. Travelling with me in vehicle six was the deputy ambassador and a diplomat called Kiko San. Both were on their phones, probably briefing worried family members about the escalating situation.

Then ops gave the command we were all waiting for. 'Gunship is locked on, H. You're good to move.'

'Let the fun begin,' I muttered as the six-vehicle convoy roared into life.

We wound our way past the security barriers at the front gate and then onto a 550-yard stretch of tarmac linking to a T-junction on the main road. But just as the lead vehicle reached the junction, a prolonged burst of gunfire sent people scattering right in front of us.

'Stop! Stop! H, we have hostiles both sides ahead.' It was the driver of vehicle one.

Shit!

'Six is pushing forward,' I shouted into the radio, steering to the left of the convoy which was now stationary. As I drew level with vehicle one, I saw what the driver was talking about. On the left, men were shooting up the road we needed to use, while on the right, a white pick-up with a PKM mounted on the back returned the fire. I could not positively identify either group, but if we proceeded, we would be caught in a deadly cross-fire.

I took a moment, maybe two seconds, to gather my thoughts. Even though our vehicles were armoured, the engine blocks were not. If a round hit one, the convoy would be instantly immobilised, with highly vulnerable passengers stranded right in the middle of a firefight.

'Abort!' I shouted into the radio. 'Reverse, reverse!'

As we sped backwards from the ongoing gun-fight, I gave a quick countdown, 'Left hand down in three ... two ... one.'

All drivers did a simultaneous two-point turn, smoothly peeling off to face in the direction we had come from, and revved the vehicles back into the compound. My first attempt at an extraction had failed miserably. But I was absolutely certain I had made the right decision. No question about it.

I turned to the two diplomats in the back. 'You guys okay?'

They slowly nodded, eyes as wide as saucers.

'It's looking decidedly unhealthy on that route, so we will reassess later today,' I said.

There was no reply. It seemed as though shock was setting in. It's easy to forget that these people had never experienced anything remotely like this before and were understandably terrified.

'We will let everything calm down,' I said, as reassuringly as I could. 'It will be better to move at night anyway, as it's much quieter.'

This was not strictly true, but I needed to put a positive spin on the worsening situation, despite the fact that I hated night moves. We had night-vision goggles, which I doubted the Taliban had, but even so, six white SUVs with bright headlights would be sitting ducks. It was a risk we would have to take.

I then sent two guys onto the roof to monitor the fighting on the road. If it died down, we needed to seize the moment and try again.

Kiko San approached me. 'H, I hope I can see my daughter again one day.'

She was definitely in shock. I put my hand on her shoulder and said, 'I will get you to the airport. Do not worry.'

She stared at me, eyes still blank.

'Ma'am, go and grab a coffee, and we will be out of here before you know it,' I said. 'It will be nice and quiet later on, and we will get you on a plane home. Let me do the worrying about that part. Just don't unpack anything, as I know how long you ladies take with your bags when going on holiday!'

I threw in the bit of light-hearted banter and got the desired smile in response. Great, I still had her with me.

Her phone rang and she scurried off to update whoever had rung her.

As she did so, a fancy-looking helicopter clattered past in the sky above. It seemed someone was leaving in a hurry. Some lucky VIP with connections, no doubt, and I briefly wondered who he or she was.

I turned to the rest of the team. 'Coffee, lads?'

We headed towards the kitchen. There were still a few hours to sunset, and no doubt we were in for another long night. Even insipid instant caffeine would be king.

My phone pinged with a message. Planes ferrying the British and Canadian embassy lads and their clients had just taken off, some bound for Kuwait and others for Oman. While I was glad they had made it out, it was strange reading the text while stuck in the Red Zone with the enemy advancing and clients yet to be extracted. For the first time I felt nervous. Nothing serious, nothing crazy. Just the realisation that a lot of people were depending on me and my team.

We gathered in the car park as evening fell, an hour before we were due to run the gauntlet to the airport for the second time. I could tell by the haggard faces, mine included, that not many people had slept while waiting for dark. We were all too hyped up, and the endless hanging around certainly didn't help. I taped a strobe light onto the roof of my vehicle, so our Apache escort could identify us, and not long afterwards the gunship appeared.

A voice from the ops room came through my VHF radio. 'H, you're clear.'

I flicked across to the team's personal channel so only they could hear me. 'Seconds out. Round two, gents.'

The first vehicle reached the T-junction and gave me the news I wanted. 'Junction's clear.' The game was on and as we turned onto the main road, I noticed some people milling around at the corner. They didn't seem armed, but then I saw an RPG leaning against a wall. Very few government troops carry RPGs, especially in built-up areas like Kabul. So that could only mean one thing. I grabbed the radio transmitter.

'Think that might have been a Tali checkpoint we've just rolled through, lads. Keep driving.'

The deputy ambassador leant forward. 'It was a Taliban checkpoint?'

'Possibly, sir.'

We pushed on in the darkness. The gunship hovered above us, but I doubted the pilot could positively ID the people manning the checkpoint or have authority to open fire. Anxiously, I scanned the road ahead. Between us and the airport lay more checkpoints, and who knew who controlled what? Afghan army? Police? Or Taliban? After seeing the RPG at the T-Junction, I decided that we would just drive for any barriers at speed and hope for the best.

At the next checkpoint a couple of people attempted to stop us, but as no one jumped in front or fired at us, I instructed all drivers to zoom through.

We then approached the bridge leading to the airport. As we got closer, I saw the distinctive camouflage uniform of the Afghan army. It seemed that the bridge was the dividing line between the two factions, and it was probably what the shoot-out had been about earlier that day. The Taliban controlled the junction and the main road, while the army still held the bridge leading to the Taipan Gate on the east side of the airport.

However, it was pointless worrying about that now. We were committed, no matter who controlled what. It was just a case of keeping a foot on the accelerator.

I had planned the route to reach the Taipan Gate at exactly 2 a.m., and so far we were on schedule. Then out of the eerie darkness, the gate loomed. We had arrived.

Grabbing the radio, I called the ops room at HQ so they could verify our call sign of six diplomatic vehicles to the Americans guarding the gate. Soon afterwards four heavily armed Delta Force guys swung the large double-steel gate open and beckoned us in.

'You the embassy people?' one asked.

'Oh yes.'

'Ramp 5, bro.'

I knew exactly where it was, as several months ago we had taken the coffin of Dr Nakamura there. As the gates clanged shut, I exhaled a massive sigh of relief. So did the diplomats in the back. Finally, we were inside the safety of the airport.

We stopped at the NATO-controlled airhead and the air-asset commander on duty that night turned out to be an RAF guy. It was nice to hear a familiar accent.

'Mate, I have the twelve diplomats for your 04.20,' I said, pointing to the convoy.

The diplomats hurriedly got out of the vehicles. I had been able to keep my promise to Kiko San.

She would see her daughter again. She turned and waved as she disappeared into the processing centre.

The RAF officer watched them go.

'Only twelve?' he asked.

'That's all the diplomats I have, mate.'

Maybe he was expecting more. I paused, wondering what he was getting at.

'What about you guys?' he asked.

Then it dawned. He thought we were also flying out.

I shook my head. 'We have to go back. Got seventy-two Gurkhas and ops guys still at the HQ, plus several hundred other people.'

'Seriously? You're going back out there? Into that shithole?'

His face said it all. We must have looked completely crazy to him. I understood what he was getting at, but the thought of not going back to fetch the rest of our people hadn't crossed my mind – or any other team member's, for that matter.

'Save me a window seat, mate. And I take white, not red wine,' I said as we sped off. I was only half-joking, as I really could have used a drink.

We took the same route back, the logic being that if they hadn't opened fire last time, they wouldn't do so this time.

Wrong call. As we approached the first checkpoint, a volley of AK-47 rounds cracked like sonic hornets above our vehicles. The bullets were so close that I knew they were meant for us.

I grabbed the radio. 'Lads, these guys are just firing warning shots for now. Keep driving.'

In my mind I kept repeating my favourite phrase: 'Movement is life.' Keep moving. That was what had kept me alive in Iraq.

We sped through the checkpoint and I noticed that the men there also had RPGs. It seemed that the Taliban now controlled the streets. Government forces were obviously in full flight.

We urgently needed to radio the ops guys and Gurkhas, who must have been going frantic while waiting for us and hearing all the checkpoint gunfire.

'Zero, this is Kilo 2 now at Yellow 32,' I said, updating the route via VHF as we rolled through our designated checkpoints, each with a colour and number assigned. Usually these were prominent roundabouts or junctions.

'Roger that.' Ops confirmed they now knew our location.

I nervously watched the GPS count down the distance back to HQ, almost in slow motion. As we got closer, I called ahead to make sure the gate would be open, as it was possible we would come in 'hot'. Whoever had been shooting at us might be following.

We zoomed into the parking lot and spun the vehicles around to face the gate, ready for instant use, if necessary. I jumped out and checked the fleet for damage. Thankfully, there were no leaks or bullet-holes. As I suspected, the Taliban had fired warning shots. Maybe they weren't sure who we were, or maybe the diplomatic plates made them hesitate. Either way, I knew we had not seen the last of them.

I gathered the team together in front of the cars for a quick debrief. We all knew the situation was deteriorating by the minute and we would be in for a long night.

'Keep everything to hand, guys,' I said, referring to our kit and weapons. 'If anything happens, it will happen fast.'

I headed back to my room, dumping my heavy gear onto the floor where I could easily find it, even in the dark.

A piercing alarm suddenly went off, the shrill siren sending shivers down my spine. It was quickly followed by an announcement on the tannoy system.

'Local guards have abandoned the gate. They say Taliban are inbound. Embassy lads, stand by.'

My team was referred to as the embassy lads, as our key function had been to protect the diplomats. We were also by miles the most experienced combat veterans in the compound. I grabbed the kit I had dumped on the floor only moments before. Snatching my weapons from the table, I raced up the stairs onto the roof, bumping into three other team members doing exactly the same. We all knew the drill like clockwork.

'Fuck me, H, this ain't good,' said one.

'They followed us back from the checkpoint,' said another.

Each team member had his own opinion as to what was happening, but I said nothing. I first needed to know exactly who we were up against. Was it Taliban insurgents, or ISIS prisoners who had broken out of the abandoned neighbouring jail, who were about to attack us? I needed clarification before I could make a decision, as these were two very different scenarios. If it was ISIS, then the numbers would be fairly small, perhaps no more than ten or so. However, if it was the Taliban, the twelve of us would be greatly outnumbered. No matter how many of them we took out, it would be only a matter of time before we ran out of ammo and were killed.

By then, the whole team was on the rooftop. I could identify each one in the dark by their silhouettes, the way they stood or the kit they had on. They quickly gathered around me, breathing heavily.

'Lads, we hold this building as long as we can. Any casualties, and Jay is going to get you off the roof and patch you up on the stairs.'

Our medic nodded, eyes glistening in the darkness – a silent acceptance that it was about to get real.

'If they use mortars or rockets to get into the stairwell, we slow them down and use the gate as a chokepoint to channel them into the compound. Single shots only, as we need to save ammo. And watch your back – this could be a diversion as they try to outflank us.'

I wanted to say more but, with the adrenaline pumping and my heart smashing against the inside of my body armour, talking was making me breathless.

'Questions, boys?'

It was the best I could do in that hopeless moment, but even so, a collective brotherhood cloaked us. We knew what was about to happen. We would not go quietly into the night.

'Fuck it, let's do it,' said Nath. A short, stocky guy with a dry sense of humour, it was his first hostile environment as a contractor, but he was acting like a seasoned professional.

We got into our firing positions. That's when I heard it – the engines of half a dozen Toyota pick-ups loaded with fighters roaring along the same road we had driven up ten minutes earlier. Through my night-vision scope, I picked out the Taliban white flag, with its black Arabic lettering, waving high on the back of one of the trucks. My already surging adrenaline levels kicked up yet another notch and my mouth went dry. The feeling of dread was unlike anything I had experienced before. Sure, I had been put through the wringer more than a few times in my life, but in the Marines and even in Iraq, we always had support.

Here, we had none. It was just me and the team. We were on our own.

I breathed in deeply, then out, slowly, willing my body to relax. These were my guys. My friends. My brothers in arms.

Twelve guys surrounded by an entire city full of Taliban fighters.

23

Anarchy, Afghan-style

My only thoughts were of survival. I had never been at such a tactical disadvantage in my life, ridiculously outnumbered and completely outgunned.

Should the Taliban choose to use indirect fire, it would be a very short firefight. The weapons we had would allow us a fairly good range to try to keep them at bay, but I was most concerned about how quickly they would decide to change tactics and bombard us with mortars and RPGs. And would they attack on all fronts? If so, we would only have three guys defending each side of the building.

'Here we go,' said Bry as he knelt, cradling his belt-fed machine gun aimed directly at the front gate. Ben and I would also cover the gate, while I placed the rest of the team in a circle to watch for anyone attempting to sneak around the other sides of the HQ premises. Arriving at the front gate could have been a diversionary tactic.

The lead Toyota slowed as it approached. In the back, I counted five armed Taliban, but I couldn't see how many were in the cab, as the glare of the headlights blinded me. Two or three perhaps. A couple of men got off the back of the pick-up, shouting in Pashto as they rattled the gates.

'The fuckers are coming in,' I said, flicking the safety catch off my M4. I was squinting through an EOTech sight, just a red dot that glowed as I tracked the front group. They then stalled. Maybe they were unsure of who we were – or more likely, they were waiting for back-up.

Even though everyone on the roof knew there was no way we were going to win this fight, staring at the Taliban through our rifle sights, waiting for everything to kick off, was excruciating. It was like being in a queue for a rollercoaster ride, but this was a ride with a one-way ticket. I just wished they would get it over with.

'Jesus, this is one mind-fuck,' whispered Ben. I couldn't have agreed more.

Out of the darkness, Ian, a loyal and reliable member of the team, rolled away from his firing position.

'Anyone want a water?'

He ripped open the plastic on a crate and threw bottles across the roof to us.

'Stay hydrated, boys.' We laughed, more from nerves than anything, but it broke the tension.

'Looks like we pull another all-nighter, H,' said Trev. He had only joined the team a couple of weeks previously, and was holding up well.

'Just make sure you put in for all this overtime,' I replied. Being on a day rate, we chuckled ironically. None of us expected to be alive to collect our wages, in any event.

Another half hour passed. By then it was obvious that the Taliban gunmen were not going to breach the gate.

Instead, they had decided to park their trucks up against it as a blockade.

We started to relax. The firefight to the death we had all been certain was about to happen had been put on hold. We gathered in the middle of the roof to discuss what would happen next.

Looking at the blocked gate, Bry verbalised what we were all thinking.

'Well, we ain't driving out now, boys.'

'Just have to use a helicopter, then,' said Ben. 'Except how the fuck does one land on that?' He pointed to the pizza hut.

'Okay, lads,' I said. 'Options are we split shifts or we do a fifty-fifty up here.'

The decision for guard duty was unanimous. Split shifts.

'Okay, splits will be four guys on and eight off, two hours on the roof with four hours' rest. Any – and I mean any – movement at the gate, get me on the radio.'

The first four-man shift took up their positions, while the rest of us went to our rooms. I doubted I would be able to sleep, so didn't even take my boots off as I lay on top of the bed.

One thing was for certain. Our new best mates blocking us in would have some scores to settle. We had been fighting these guys for twenty years and bitterness, or more likely outright hatred, ran deep. They loathed foreign security contractors as much as they did foreign soldiers. There was absolutely no doubt we were the enemy. The stand-off might even end with us playing starring roles in our own movie – a snuff video. Who could know?

Being the last out of the Green Zone had undoubtedly cost us dearly. At least we had got the diplomats out. But at what price? Our lives?

Trying to sleep when surrounded by fanatical enemies is like attempting to nod off during a parachute jump. Not possible. Whenever I closed my eyes, my imagination ran riot picturing what could happen next. Would we be publicly executed, held hostage and tortured, or just abandoned and forgotten? What would my kids think in years to come about growing up without a dad? And what would I have died for?

Nicky would have to explain that to them. A truly horrible thought and a massive reality check for me.

I eventually gave up trying to sleep, put on some music and made myself a coffee. A couple of slugs of the instant stuff and my night just got worse.

As dawn broke, I was on the roof with Ben, taking the last watch. Daylight revealed our new friends in full colour rather than the murky green of night-vision optics. Two trucks remained at the gate, with several heavily bearded and even more heavily armed Taliban controlling the road outside: the only way to the airport and freedom.

'Shall we ask them if they want a bacon sandwich?' Ben asked, knowing full well the Taliban didn't eat pork. 'Or take them a few beers? Must be hot in that sun all day.'

We had a good chuckle. 'Well, mate,' I said. 'I don't know about them, but I could kill a beer right now.'

An ice-cold Corona with lime would be heaven. Even though it was six in the morning, I felt we had earned one.

I only kept a handful of guys up on the roof during daylight. There was no rigid routine like the night before, and if the lads wanted to make phone calls or a coffee, then they could. I also told everyone to keep their weapons out of sight. Armed people on the roof could provoke the Taliban and the last thing I wanted was an unnecessary gun-fight.

I noticed there were quite a few Eastern Europeans outside, protecting their accommodation premises. These guys had been employed by HQ on static security contracts – gate or checkpoint guards – and I had not seen them last night. Maybe I needed to bring them into my far more competent combat team to pump up our numbers a bit. Not that it would make much difference, as with the Taliban rapidly taking control, we would literally be fighting against the whole of Afghanistan.

We then learnt that our diplomats were not the only people getting out of the country. The day before, President Ashraf Ghani had fled for Dubai, but not before allegedly loading his helicopter with US$169 million. I now knew whom that fancy helo flying over HQ belonged to. I shook my head at such cowardice. What a craven way to show leadership. No wonder his army was sprinting for the hills.

Jay, our medic, then came onto the roof. 'H, bit of a problem, mate. Some of those Gurkhas are properly flapping.'

I knew what he was talking about. We had noticed that whenever we moved anywhere, even simply changing shifts, the Ghurkhas would rush up, asking if we were leaving. They obviously thought they were going to be left behind.

I was not in charge of the Gurkhas, but something had to be done. I walked down to their floor just below ours in building five and called over the senior Gurkha, whom

I knew. He had been head of night-shift security back at the embassy.

'Has anyone told you what's happening?' I asked.

'No, sir.'

I couldn't believe it. These guys, for whom English is mostly a second language, had been left completely in the dark. No wonder they thought they were going to be abandoned.

'Okay, call your men for a briefing,' I said.

They gathered around and I painted the picture as brightly as I could. Who knew what was going to happen next anyway?

I said the ball was now in the Taliban's court, as it appeared they had decided to block the HQ gates for the moment. This meant we could not leave for the airport just yet, so we had to sit tight and wait. We could possibly get airlifted out, but first we would have to dismantle the pizza shack. I stressed they would not be left behind under any circumstances. Not on my watch.

As I was speaking, the radio clipped onto my belt came to life. 'H, mate, you need to come onto the roof of building five and see this.'

I ran up the stairs, wishing I had a quid for every flight I had gone up or down.

'What have you got?' I asked.

One of my guys handed me a pair of binoculars and pointed in the direction of the airport.

'Crowds of people are now at the North Gate and getting pretty rowdy. Apparently, the Baron is the same.'

The Baron Gate is named after the Baron Hotel, a 160-room luxury establishment just 200 yards from the airport's boundary and an evacuation spot mainly used by the US military. I knew quite a few lads who had gone out via the Baron during the past few weeks and it had been very busy even then. Now it was completely swamped. With the imminent fall of the capital, thousands of people were fleeing for the airport, as that was the landlocked city's only exit point. Most of them had no travel documentation, and even if they had, all commercial flights had stopped. Only military planes were flying. What also worried me was that the plane with our diplomats had still not taken off. If the airport was breached, who knew what would happen to them.

I watched for a good fifteen minutes. Our worst nightmares were coming true faster than a speeding bullet. Civil unrest was flaring like an inferno, fuelled by uncontrolled panic. Many Afghans vividly remembered the brutality of the Taliban, who had first taken over the country in 1996, and knew what horrors to expect. Even worse fates awaited anyone who had worked for the Afghan government or the Americans. So the panicked mobs trying to get into the airport were to be expected. Despite the Americans trying to put a positive spin on it, the situation was eerily reminiscent of the chaotic final withdrawal from Vietnam in 1975.

'Well, that's a clusterfuck,' I said, handing back the binoculars.

A few hours later, word came through that the diplomats were wheels-up and out of Afghanistan. It was the news my team had been waiting for. Our part of the deal had been fulfilled. We had done our jobs. Now it was time to focus on how to get my guys and the rest of the people at HQ out.

This was going to be a horror show, no matter which way I looked at it. There were more than 400 people in the compound. The high number was not just Eastern Europeans and Gurkha static security guards. To my shock, I suddenly noticed that quite a few foreigners had also come in.

It was bizarre. I had no idea who most of these people were. On one occasion I saw a French doctor, two medical students and an English literature teacher from an international educational programme standing with their bags at the helicopter landing site. I asked the doctor what they were waiting for, hoping it was a pizza rather than a helo.

'Oh, we have a helicopter that's taking us to the airport,' she replied. 'Then we are going out on a charter flight.'

I was gobsmacked. No charter flights were leaving Kabul – nor could any helo land on the pizza hut.

When I told her that, as gently as I could, she was astonished. 'But we paid for it,' she exclaimed.

I walked off, even more confused. These terrified people claimed to have paid for a private charter package that did not exist. I have no idea if that was true, or to whom they allegedly paid the money. All I knew was that my team and I would have to get them out. As if we didn't have enough hassles.

However, I kept that shocking news to myself. My guys were already pissed off enough and needed to focus on the task ahead.

My phone beeped twice with a message from a specific WhatsApp group chat of intel guys who had good sources.

'H, that's the Yank Embassy all out and gone.'

So the American Embassy had gapped it. We were officially the last group of non-military foreigners in the city, although many desperate individuals were possibly still caught up in the mayhem.

'How's it looking at the airport?' I typed.

I waited a few seconds, hoping for some good news.

'It's carnage down here at the moment. You guys good?'

I gave him a thumbs-up emoji, although that certainly did not reflect my feelings. Instead, I was consumed by anxiety. 'Carnage' – even though my source meant it in the chaotic rather than massacre sense – was not the word I wanted to hear. What would happen to all the people relying on me?

I physically forced myself to get a grip. These worst-case what-if scenarios were crippling me and I needed to get myself under control. As long as I and my team held it together, no matter what happened, we could say we did the best job we could with the terrible cards we had been dealt. I had to rid my mind of all negative thoughts. I breathed in deeply. Then out. I started feeling better and vowed nothing would get me down from then on.

I went off to eat, not that I was hungry, but I hadn't had any grub for two days. Being a bit of a health freak, I don't normally do junk food but this time I loaded my plate with chips and a burger. It seemed pointless eating healthy food, as every meal could be our last.

One of the team at the table told me he had heard that some contractors who had just arrived back in the UK had had to pay £3,000 to go into isolation in hotels as a Covid precaution. I chuckled at the irony of having to spend top-rate danger pay on being quarantined.

Then another guy looking at the news on his mobile phone suddenly exclaimed, 'Fucking hell. Check this!'

The Taliban had just seized the presidential palace. A group of bearded men in tunic shirts and baggy pants were sitting at the president's desk, declaring total victory and a new radical fundamentalist government. They had complete control.

We watched the news clip silently. Yesterday seemed like a lifetime ago. We were now, for all intents and purposes, little more than prisoners hemmed in by Taliban trucks.

As that video went viral, rampant chaos and gridlock further exploded, as it appeared that everyone in Kabul who did not support the Taliban was trying to flee. From the rooftops we monitored the absolute anarchy at the airport. All entrances, manned predominantly by American Marines and special forces, were now closed and there was little chance that they'd be reopened. We heard regular warning shots being fired, an indication of how perilously close the panicked crowds were to breaching the gates. How the hell would we get through all that? Equally importantly, would the Taliban even allow us to leave? There seemed to be no way out.

As the afternoon sun started to slide behind the Hindu Kush, we took up positions on the rooftop once again. I figured that if the Taliban were going to attack, they would do so under cover of darkness. That's what I would have done.

I took the first shift on the roof with Ben. We discussed our options, even though by any sane reasoning this job was well and truly screwed. We also joked about all the text messages we were getting from worried family and friends, urging us to get out of Kabul ASAP and to 'stay safe'.

No shit! Yes, we needed to get out yesterday. And no, we couldn't stay safe.

Dark humour is the best medicine. We started laughing. I came off the roof in a much better mood than when I had gone up. I even slept for two hours.

Then I got a call. 'H, we have movement at the gate.'

I grabbed my kit and rifle, sprinting up the stairs onto the roof.

'What've we got, boys?'

'Taliban are in the compound, mate. They've breached the front gate.'

24

Dancing with the Taliban

My guy didn't even look away from the NVG as he briefed me. I could see his eyeballs lit up in green.

'How did that happen?' I asked. 'The outer gate was meant to be locked.'

A few guys chuckled and I realised what I had just said. In our opinion, the compound's security system was not exactly top notch. But it was too late to worry about that now.

'Tali's now inbound to the inner gate. Single intruder.'

The four of us on the rooftop slowly crept over to firing positions. The last thing we wanted was to spook the guy inside the premises. Instead, I wanted to see what he was going to do. Would he open the inner gate for a full-on attack? Was he planting explosive charges?

Trev leant forward and whispered, 'You want the rest of the guys up here, H?'

It was a little before midnight, so most of them had already done their shift and would be asleep.

I needed to think.

'H?'

'Yeah, mate. Gimme a sec.'

I considered using only the four of us, but I needed guys on building six to cover the gate to the main compound.

'Yup, get them up onto roof six, mate. Tell them to move around the back of the building and up the stairs. But very slowly.'

I wanted to avoid any knee-jerk reactions. If the insurgent suddenly saw ten big, heavily armed guys running up to a roof, he would assume it was a hostile act.

Anxiety and dread gushed through me. Was this it? Would it kick off now?

We watched the intruder as he checked the inner gate. After about twenty minutes, he went back to the main gate where his mates were.

'Okay, stand down, lads.'

The next morning, the commander of the Taliban unit contacted HQ management and said he wanted a meeting that night. It was then that we discovered why the fighter had come into the premises. Compound guards had neglected to lock the main gate, and the Taliban wanted to know why it was open. Maybe they thought somebody had tried to escape. I don't know, but the irony of our enemies asking us why our gates were open pretty much summed up the crazy situation for me.

Even worse, if we had panicked and fired at him, we probably would no longer be alive.

The morning dragged on. Gunfire continuously crackled from the airport, and from the roof of building five we watched anxiously as the situation deteriorated. The airport was our last hope; it simply could not fall. Everyone in the HQ compound knew that.

Then Nath held up his phone. 'Just heard that the airport's military side has been breached.'

'Jesus ... seriously?' asked Trev.

There was a moment's silence. 'Well,' said Nath. 'If we weren't fucked before, we most definitely are now.'

We huddled around Nath's phone, watching videos of people fighting their way through the North Gate and stampeding onto the runway.

'Fuck! Look at that!'

We watched in shocked fascination as a US military plane taxied down the runway, followed by sprinting Afghans, and eventually took off with people desperately trying to cling onto the undercarriage. The video ended with two bodies plummeting like sacks of stones as the plane climbed into the sky. For several moments, I could not speak. None of us could.

Those awful scenes would soon be screened around the clock on TV news and I felt sick knowing that Nicky would see them. I could no longer play down the actual horror we were facing. She wasn't stupid, and I dreaded having to update her later. The last thing I wanted was to put her through more pain and worry.

Trying to erase that harrowing video from my mind, I instead focused on the meeting later with the Taliban. As my guys would have to provide security for it, I decided Ben and I would be present in the meeting room, while the rest of the team would be in firing positions on the roofs of buildings five and six. I grabbed some old AK-47s and spare ammo, strategically placing them along the route from the meeting room to the rooftop. If everything went loud, Ben and I could grab them and shoot our way upwards, securing the rooftop entrance by bolting closed two heavy steel doors. From there we could defend ourselves from high ground, where the rest of the team was, and Bry, our man-mountain, had two belt-fed PKM 7.62mm machine guns. We could buy some time with that, but not much else.

With our worst-case-scenario strategy in place, it was time to sit back and wait for the sun to set on another nerve-racking day in Afghanistan. We had a meal as a team just before last light, but I barely ate anything. It was more a chance to sit with the guys and see how everyone was holding up. Morale was pretty good, despite the impending visit of tonight's 'special guests', which showed the quality of men I was with.

An hour before the scheduled meeting, I put two guys on building five, with the rest on six. Ben and I would be armed and stick close to the four unarmed management guys. While we were waiting in the corridor, one HQ staff member started pacing up and down, clearly agitated. 'I'm telling you now: this is an ambush,' he kept repeating. 'They're going to come in and wipe us out, I'm telling you!'

Ben and I looked at each other quizzically.

'Jesus, he's lost it,' Ben whispered. I nodded. This was the last thing we needed – someone having a panic attack. Fortunately, he disappeared down the corridor before the Taliban arrived.

Our radio crackled into life.

'HQ foxtrot to the main gate, H.' Foxtrot was military radio jargon for 'moving'.

'Roger.'

HQ management was walking out to meet the Taliban. My guys would cover every movement from building five. It was a tense few moments until I got the next transmission.

'H, that's the guests now picked up. Seven in total, all foxtrot to you.'

They were coming into the compound. I clicked the transmitter button twice to confirm. No need to speak, as I didn't want to interrupt any radio intel from my spotter on the roof.

It came right away.

'H, these guys are carrying. So far I count four armed.'

'Roger that. Four armed.'

Ben and I were already outnumbered. We would need to take out two each.

The roof spotter started a countdown. 'H, ETA to you thirty seconds.'

'Twenty'

'Ten.'

'Now!'

I came face to face with the Taliban. Armed, dangerous and intensely suspicious of us. The feeling was mutual. Making our way into the room, I leant close to Ben.

'I have Tubs, blue shirt and grandad.'

I picked my targets quickly, calling out easily distinguishable features – shirt colour, age and build – so he knew exactly who I would take out. He would gun down the rest if everything went south.

The atmosphere was tense, which was hardly unexpected. However, I tried my best to appear as relaxed as humanly possible, smiling as if I had met them a hundred times before. Nicky always said I had the gift of the gab; let's hope she was right.

The Taliban and HQ management sat down as Ben and I stepped outside the room, leaving the door slightly ajar so we could burst back in if we had to.

The meeting started via an interpreter, while we paced the corridor. After twenty minutes we heard chairs scraping, an indication that the talks were over. I opened the door.

'We're headed to the underground car park, lads,' one of the HQ guys said. I nodded and relayed the news to my team.

I knew exactly why they wanted to go to the undercover parking lot. That was where HQ staff had stored all the weapons. I had watched throughout the day as piles of pistols and assault rifles, as well as operational kit ranging from satellite phones to body armour, were simply dumped on the floor. The basement was a war-fighter's dream.

Ben went ahead and I slotted into the rear, trying to control the formation of the group. Our management guys should have stuck together, but for some reason they mingled with the Taliban, making it difficult for us to get clear shots if anything were to happen.

The Taliban were overjoyed when they saw the kit piled in the parking lot. They were like kids opening presents at Christmas – I don't think I have ever seen such excitement. They couldn't get among all that gear fast enough. However, the distraction gave me an opportunity to surreptitiously reposition the HQ guys, so me and Ben would have a better line of fire if we needed to shoot our way out of this.

After a few moments the guy I had nicknamed Tubs came over. He was all smiles, blissfully unaware that I had earmarked him as one of my targets in the meeting room. He pointed to my weapon sling and then to his own AK-47. From what I could gather, he wanted my rifle sling. The problem was that I had an AR-15 and his AK wasn't compatible. Also, I wasn't about to start giving my kit away.

I walked over to a box of radios placed on top of some slings off other weapons. I grabbed him one. 'Here, mate.'

Tubs was beaming from ear to ear. My new best friend, and hopefully that would earn me some brownie points if I found myself staring down the wrong end of his AK.

Unfortunately, he didn't know how to put it on and indicated I must do it for him. So we ended up in a manoeuvre that resembled an awkward teenage bop at some town-hall disco. I needed his weapon to fit on the sling, but I was holding my own. Reluctantly, I let go of my rifle, and pointed to his AK. He gave me a sideways look, then handed it over. His safety was on, but I wasn't sure whether he had a round-chambered rifle, so I only held the stock and the barrel as I fitted his sling through the loops and tightened the fixing. I then held it up so he could slip his head underneath.

I couldn't help smiling at the crazy situation. I was basically dressing this Taliban guy for battle – hopefully not with me, but you never knew.

I thought that was the end of that, but I was wrong. Tubs started rabbiting on in Pashto with a mate, and when the conversation stopped, they both looked at me. He then walked towards me, beckoning the interpreter over and gesturing to his AK. What had I done wrong? My finger moved surreptitiously to the AR's trigger.

I waited for the translation. When it came, I almost burst out laughing.

'He wants you to make his sling look like how the Russians have theirs.'

How the hell would I know that? The Russians had invaded Afghanistan in 1980, when I was barely a toddler. My mind raced back to all the documentaries I had watched on the mujahidin fighting the Soviets, but only a complete geek would remember something as obscure as a rifle sling.

I decided to blag it and made it into a version of what Tubs looked like – sort of droopy. As I handed it back, I could see the disappointment in his eyes. I no longer rated my chances if I found myself at the opposite end of his AK.

With prized new war-fighting goodies in hand, the Taliban turned their attention to the vehicles, counting each one and asking who they belonged to. Happy with the answers, they then turned to go.

The mood was almost jovial as we shook hands at the gate and Tubs even gave me another grin.

Perhaps that was just because he knew they were going to attack later that night and he could give his brand-new sling a test-drive. I hoped not.

Ben and I returned to building five and the team gathered around for a debrief. I told them about my comical dance with Tubs, and then said it would be a good idea to start ditching personal kit like watches and rings. My team had a bit of a thing for expensive watches, and an Omega Seamaster or a Breitling was a pretty big payday for these boys. Certainly more than a rifle sling, droopy, Soviet-style or otherwise.

'For now, lads, it's a wait-out.'

I was growing bored with hearing myself say that, and I knew the team was, too, but there was nothing more we could do.

The next morning at breakfast, Rik, a team member who had been in Iraq at the same time as me, said, 'Lads, did you get that intel from ops?'

'Not seen it yet,' I replied.

'The Taliban guy who headed up that meeting last night is the same one who planned the Taverna du Liban Lebanese restaurant attack.'

I sat up, staring at Rik. This was bombshell news. The Taverna du Liban had been extremely popular with Westerners and on 17 January 2014 a Taliban suicide squad stormed the building, killing twenty-one people, including Wabel Abdallah, the head of Afghanistan's International Monetary Fund. It had been one of the most high-profile attacks at the time.

All of us silently digested the information. Without knowing it, barely twelve hours earlier we had been negotiating with someone who had once been one of the country's most wanted killers. So what was such a big shot doing here? Was he planning a similar attack on HQ and had he used last night's visit to check the place out? If so, he would now know beyond doubt how undermanned we were.

Obviously, none of us had the answers, but for Rik it was also personal. His best friend had been killed in that attack. Breakfast suddenly became unappealing for all of us. I pushed the eggs and hash browns around my plate and tried to digest the really shit coffee.

We trudged silently back to building five and one of the ops team met us on the landing.

'Lads, I know you've probably heard that the Taliban commander here was the Lebanese restaurant mastermind. But there's more.'

Everyone looked up. He certainly had our attention.

'Three of the Taliban that came in last night were wearing suicide vests. It's just been confirmed.'

My stomach lurched. Ben and I had been right up close and personal with them in the garages, even sorting out the whole ludicrous sling saga. Had I been 'dancing' with a suicide bomber?

The bottom line was that the Taliban certainly meant us harm. At the very least, we had been dealing with a death squad and our whole group could have been blown to smithereens. There would have been nothing Ben or I could have done about it. In all my plans, such as having loaded rifles at strategic positions for a fight to the rooftop, I never for a moment thought to factor in suicide bombers. If that had happened, we would not have got anywhere near the escape route.

Rik had retired to his room, so I popped in to see how he was. He was a veteran operator, and had done and seen as much as I had, but to come face to face with the guy responsible for killing your best mate – well, it had to hurt. Sure enough, I could see just by looking at him that revenge was uppermost in his mind. He was seething with anger. We had a quick chat and I stressed that our own survival, rather than personal grudges, was paramount. We simply did not have the luxury of mourning the past, and at that moment any form of payback was not an option. It would get us all killed.

Rik was a good and seasoned operator; he agreed that he would stay cool and calm. I told him I understood and appreciated exactly how hard that was for him.

I headed down to my own room and made another coffee. What more could go wrong? We were surrounded by the Taliban, our vehicles were blocked in, the airport was in total chaos, with people falling off planes – and the commander we were negotiating with was responsible for killing a team member's best friend.

On top of that, three of his henchmen had come into our compound wearing suicide vests, proving beyond doubt that they had evil intent.

I sat on my bed, trying to absorb it all. How much more of a clusterfuck could we be in?

Something I have always been keenly interested in is polar exploration, particularly the early heroic trips to Antarctica. Among my favourites is Sir Ernest Shackleton, whose ship became trapped in sea-ice and eventually sank. His men survived a brutal winter on drifting ice floes, after which Shackleton and five others sailed an open boat through the roaring forties, the planet's most dangerous seas, to get help. It's an unsurpassed story of grit and determination.

That's exactly what I needed at that moment.

25

Mind Games

I perked up a bit later that day. Or perhaps I should say that a certain member of my team perked everyone up.

He had earlier snuck down to HQ management's office and 'acquired' a state-of-the-art coffee machine. Fortuitously, one of our lads also had a whole bag of Black Rifle Coffee Company beans that he had brought out in his embassy extraction kit. It was a true lifesaver and the team gathered around for the first freshly brewed real coffee in four days. What a time to be alive!

We kicked back, sipping caffeine strong enough to float a bullet, and shot the breeze. Coffee lifted everyone's mood, and soon we were all laughing and joking. The two guys on the roof got a cup each as well, so we were all back in the fight, physically and – even more crucially – mentally rejuvenated.

Late in the afternoon, word came down that we would be moving out that night. The Taliban had brought in a number of coaches and it seemed they might honour their side of the bargain in getting everyone in the compound to the airport. I wasn't convinced, so decided to have a Plan B ready as a base for my team to work off.

There was a whiteboard in the corridor and as the lads prepped round two of life-saving coffees, I began to draw up how I saw our extraction plan working. In total, I had six SUVs and two minibuses to move the seventy-two remaining embassy staff in one trip – in other words, my team, ops room staff and the Gurkha guards. I crunched all the numbers and assigned each individual a specific seat. Those who did not fit into the minibuses would squeeze into the back seats and boot space of the SUVs. Two of my team would be in each vehicle: a commander and a driver. It would be their responsibility to check that everyone was present with a valid passport.

HQ staff – including non-compound civilians, such as the French doctor and her students, who had claimed to have paid for a charter flight – would go in the Taliban coaches.

I opened up the floor to questions, and the main one was what to do if the Taliban reneged on the deal and attacked us. Well ... our options were limited, to put it politely. However, each SUV had a weapons system and a radio, while spare ammo was stored in a deliberately nondescript bag. So if everything went kinetic, we could at the very least try to shoot our way to the airport, despite it being surrounded by Taliban. Desperate as that sounds, I knew first-hand from my time in Iraq exactly how much incredible punishment these armoured SUVs could take. As long as the engine was good, it was possible to drive out of the worst situations.

As night fell, we packed everyone into the vehicles and waited. And waited. Then, for the second time in as many days, I came face to face with the Taliban, as they escorted two coaches into the compound. I made sure to keep my distance this time, as the intel about the previous group wearing suicide vests had really spooked me. Ominously, the mastermind of the Lebanese restaurant bombings was one of the first to get out of the coaches, barking orders to his men and scanning the buildings as he spoke. I wondered what he was thinking. Perhaps where to kick off an attack? I quickly glanced at Rik and saw that his jaw was clenched tight but his anger was under control.

The clock ticked as we waited in the vehicles, engines running, ready to roll. It was well past midnight and people started getting restless, cooped up and completely at the mercy of the heavily armed Taliban.

I then noticed that Mr Mastermind had disappeared. 'What was he up to now?' I wondered, anxiety and adrenaline coursing through me. Despite this, I grinned sardonically to myself. I worried when I saw Mr Mastermind, and worried even more when I didn't.

Soon after 2 a.m., the Taliban suddenly decided it was too dangerous to move us, saying civilians were still mobbing the airport gates. We would have to wait for the mayhem to be brought under control.

This was a massive blow, and difficult to absorb. My guys jumped out of their vehicles, slamming doors. The air was blue with profanities. The consensus was that the Taliban was deliberately messing with our minds.

I stayed in my car to collect my thoughts. I was as gutted as the next man, but losing it wasn't going to achieve anything. Controlling my fluctuating emotions was the biggest challenge for me.

A personal worry was that Nicky knew I was meant to be going to the airport that night. It was 11 p.m. UK time and I punched out a message telling her it hadn't happened.

I was aware it would cause her more pain, but I was not about to lie. In fact, I couldn't even if had wanted to, as the critical airport situation was headlining every TV news bulletin. I put as much of a positive spin on the message as I could, but I doubted that would work. It was like being in a relationship with a lie detector. We had been together for so long that she knew my mood from one text to the next.

By then everyone was back in their rooms, no doubt as totally pissed off as I was. I tried to sleep, but the moment I started drifting off, I jolted awake. I was operating way beyond normal human capacity, but fear does strange things to one's psyche. I must have been burning up energy like a racehorse and my senses were heightened to almost supernatural levels, yet I wasn't hungry or tired.

Breakfast with the team later that morning was a sullen affair. The Taliban's little mess-about with us had not been appreciated. I listened to what everyone had to say, letting them vent their frustrations, and hoped that would clear the air. It did not. If anything, it reinforced the team's perception that the situation was not being handled properly.

I said I would ask ops for an update and Rob, the popular project manager, came out to address us. A big, jovial Scotsman with a mop of thick black hair, he was an extremely likeable guy with a rare ability to put people at ease.

'Okay, lads, I appreciate last night didn't go to plan and we have to sit tight. It's looking like it's now only on tomorrow, with the airport situation deemed too dangerous to move at the moment.'

That did not go down well, and Rob got some serious verbal abuse about what he or management should be doing. The backlash reached fever pitch, but Rob kept his cool, letting everyone have their say as he tried to answer every angry outburst. It was one of the finest displays of leadership and people management I have ever witnessed – particularly in a room full of furious alpha males. With everything about to fall apart, one man managed to hold it all together.

Thanks to Rob, the team got everything off their chest and, like the true professionals that they were, hit the reset button and went back to work. If tomorrow's extraction was on – and everything in those dire days was a big 'if' – we needed to get back to the drawing board. So I set up the whiteboard and again laboured on about headcounts and passports. I couldn't stress that strongly enough. If we left one person behind or forgot a single passport, even if we made it to the airport, it would all be for nothing.

Management was due to meet the Taliban in a few hours to discuss what would happen next, perhaps even getting the green light to leave later that night, so once again we returned to wait in our rooms. These delays were absolute purgatory and doing nothing triggered more anxiety. It was impossible not to visualise worst-case scenarios. If these fundamentalist fanatics wanted revenge after two decades of fighting, our premises were the perfect killing fields. About 400 foreigners were crammed into the compound, twenty of them British, whom the Taliban hated almost as much as the Americans. For some reason I thought back to how al Qaeda in Iraq had killed the Blackwater contractors in Fallujah and hung their corpses from the bridge. Would we suffer a similar fate? Would my body be strung up from the rooftop my team and I had been guarding for the past four days and nights?

Questions, questions, questions. But never any answers.

A call came over the radio. 'Embassy lads, standby. We're meeting in figures five.'

The Taliban had arrived. Ten of my team headed up onto the roof, while Ben and I went to the meeting room. It was the same plan as before: tactically placed weapons systems made ready in case we needed to make a run for it. My guys would slam shut the big steel doors leading to the roof once we were clear, and then it would be a fight to the death.

'Foxtrot to conference room, two minutes.'

The management guys were two minutes from me and inbound with the Taliban.

'How many pax have we got, boys?' I asked.

'You have seven guests, all armed with longs.'

'Long' was a term we used for assault rifles. In this case the Taliban were carrying 7.62-calibre AK-47s or M4s taken off government troops.

'Roger.'

'Inbound in ten secs, H.'

'Here we go again,' I thought. The bearded men in baggy white trousers entered the room and nodded to me and Ben. We had no longs and had opted to hide our pistols, appearing unarmed. The Taliban seemed happy with that. I even got a smile from another fat guy looking a bit like Tubs. I smiled back, as if greeting mates at the pub. But in reality, Ben and I were quietly identifying which target each of us would take out if we had to.

As we moved into the hallway, I whispered, 'Get me the fuck out of here.'

Ben quickly retorted, 'What's worse, this or Lincoln?'

We laughed silently, shoulders heaving in the darkness of the corridor.

'Probably Lincoln, mate.'

The meeting finished and we got more bad news. We wouldn't be leaving that night, as the Taliban said they didn't have enough coaches for 400 people. Until they did, we would have to sit tight. Rob was present in the meeting so he would have to update the team and weather the resulting shitstorm. He obviously was not relishing doing that, so I tried to lighten the mood.

'Hey Rob, maybe just say that the Taliban want the lads to do weapon-handling training?'

It was an ongoing joke with the team that every month we had to practise weapon skills, despite most of us having more than twenty years' experience with every firearm imaginable.

It had the desired effect and Rob started laughing. He took the joke further: 'Taliban want you guys in shorts and T-shirts for a physical fitness test.'

'Taliban doing a locker inspection at 05.00.'

By the time we reached the accommodation, we were almost weeping. The team heard the guffaws as we came up the stairs and asked what was happening. We shared the joke and everyone saw the funny side. Yes, we were not moving, and yes, we were still surrounded ... but it could always be worse. At least we were still alive.

They took that hit like true champions. Our brand-new coffee machine, 'courtesy' of management, was fired up.

If the Taliban attacked that night, we would at least go down laughing, caffeinated and motivated.

26

Checkmate

Finally, we had a glimmer of hope. The Taliban gave instructions that everyone in the compound must be in the car park at 7 a.m. the next day.

That night's sleep was as restless and broken as ever, despite having pushed through seven days like this. I stared blankly at the ceiling, wondering what the Taliban would do this time. More mind games to further break our morale? Would they attack now they knew what weapons and how many fighting men we had? Would they keep their word and allow us to leave?

Or would this be my last morning on Earth?

I threw off the covers and got up to make a coffee. Then I sat on the bed and completed my death letter to Nicky. Sending it to her if everything went kinetic would be the last thing I did in this lifetime. It would be telling her from the depths of my soul how much she and the kids meant to me.

It was time to go downstairs. I opted to wear jeans and a grey T-shirt to look as 'civilian' as possible, as well as a Ducati baseball cap that I could pull down over my face in the event of a snuff video.

Standing there, looking like I was off to the pub, I threw a backpack over my shoulder, and dumped my M4 and pistol under the mattress. If the situation went loud, I could at least try to run back to collect my weapons for a last stand.

Stepping outside, I was instantly hit by the August heat. It was only six in the morning but the sun was already a furnace, forcing me to squint, as my sunglasses were inside my vehicle.

The six white B6s were neatly lined up and ready to roll. A few had doors open and the team were sorting out various bits of kit, hiding weapons and ammo inside their allocated rides. We exchanged nervous smiles and nods, and I noticed that the three Gurkhas travelling with me had already got into the back of the vehicle and made themselves comfy among the baggage. They were taking no chances.

'Morning, lads.' I flashed a smile and handed the Gurkhas my bag. 'Think it's about time we got out of here, hey?'

I got a chorus of nervous laughs as a reply, but I wasn't exactly expecting much in the way of conversation from anyone that morning.

At the front of the vehicle, Ben, who would be driving, had everything up and running.

'All right, mate? You sleep?' I asked.

He looked exactly how I felt. 'Nah. Well, kind of, but not for long.'

As our departure time approached, I got into the commander's seat and turned to my passengers.

'Passports, boys?' I asked. Everyone held them up. I may have been the team leader, but I held mine up as well, so we had a hundred per cent accountability check. Seven people, seven passports. Rob then came over and leant in my window. 'H, can I have a word?'

This was not going to be good. The fact he was pulling me out of the car and away from the team, as well as his edgy body language, was an indication that the situation was already going south.

'Listen, mate, the Taliban aren't going to let us take the vehicles.'

The words hit me like a sledgehammer. Without the B6s, we had lost our armour protection, as well as the weapons we had hidden inside. We would be totally at the Taliban's mercy.

'Right, so what are they suggesting? We have seventy-two people from our embassy to extract.'

I liked Rob a lot and tried my best not to sound too pissed off. But seriously, we could not have got worse news.

'They say they have extra buses coming. But they don't know when.'

Right! Another Tali psycho mind game. I glanced at the main fence where some Taliban foot soldiers were pacing up and down, staring into the compound. They reminded me of caged tigers waiting for a chance to attack vulnerable prey.

'Look, I will tell my guys. But we're massively exposed, mate.'

I could hear the worry in my own voice. No doubt Rob could, too. We had nothing to defend ourselves with. The killing fields scenario, my ultimate nightmare, loomed large in my mind.

'If anything changes, I'll let you know,' said Rob. 'For now, it's no cars.'

Hiding disappointment is not something I do well. Even less if lives are at stake. My team, who had watched the exchange with Rob, knew something bad was happening. I called them over.

'Lads, it's no cars. So, effectively, no weapons. Just body armour and helmets.'

No one said a word. Normally, there would be an angry chorus of 'bloody hells'. There was just a quiet acceptance. Shit happens.

As I spoke, we heard the growling of a diesel engine approaching. I couldn't see exactly what was happening, but it seemed as if a large coach or truck was reversing up to the main gate of the compound. At the same time, the HQ management team came out of the office and walked towards the gate, one of them carrying a large red bag that had been taped up. A Taliban soldier took the bag and opened the gates.

We all looked at each other. What the hell was this all about?

I shrugged. Whatever was happening before our eyes was between management and the Taliban, and I guess that some 'donations' might have been demanded as part of the negotiation process.

A well-dressed guy who spoke almost perfect English – the antithesis of the type-cast Taliban fighter – then entered the compound with five others, standing in a line. They must have got recent upgrades because they all carried shiny black American-issue M4 assault rifles, no doubt courtesy of the former Afghan army heading at speed for the hills. The battered AK-47 I had put a sling on earlier that week, and all the others, were history.

HQ management and the smartly dressed Taliban dude engaged in some serious conversation for three or four minutes, then the gate was closed again.

'We ain't fucking going anywhere,' Nath muttered in disgust. I shared his disappointment, pulling out my phone to make sure the final email to Nicky was still ready to go at the press of a button. It was.

Management called Rob over for a briefing. The tension of not knowing what was happening was killing me. I was done with these endless mind games.

'What the fuck is going on?' I shouted, as Rob walked back to us. When I got the answer, I wished I had never asked the question.

'No body armour, no helmets allowed,' said Rob. 'They will search everyone's bags before we leave.'

'You have got to be kidding me,' I thought. First our weapons, and now our body armour. An execution rather than an extraction was becoming more and more likely.

After we had been searched, and every weapon and piece of operational kit confiscated, we were herded into an area enclosed on three sides by walls. The perfect killing zone. At that moment two more Taliban trucks arrived with PKM machine guns mounted on the back. The trucks blocked the entrance and exit to the enclosed area. The machine guns were pointed at us.

I pulled out my phone, thumb poised to send my death letter to Nicky. This was it, the end of my life. I was at peace, knowing that my last living act would be to send a message to those I loved most.

*　*　*　*

Suddenly, there was movement ahead. The Gurkhas at the front of the queue had started getting onto a coach.

I relaxed my thumb on the phone, but still kept it close, checking to see what the gunners manning the PKMs were doing. They, too, seemed to be getting ready to move off. I watched, barely breathing, my mind still unable to accept it. After everything that had happened, I was still convinced this was a trap. They were going to load us into the coaches, and then they would open fire.

Yet the line continued to move. Dare I begin to hope?

Eventually, my team reached a minibus designated for us. We crammed into it, big guys squashed together, shoulder to shoulder. I was last in and sat in the front, next to the driver. He appeared to be a local civilian, press-ganged by the Taliban to drive the coach, and seemed as nervous as we were. I guess driving to a besieged airport with a load of foreigners wasn't his idea of fun, and I didn't blame him.

As we waited, Trev noticed his expensive ballistic helmet perched on the head of a Taliban soldier standing outside in the road. As he cursed the thief, we all burst out laughing. It was the pressure-valve release we all needed.

After some shouting in Pashto and grinding of gears, the convoy started moving. We snaked out onto the T-junction, and instead of driving straight into a gun-fight, as we had with the diplomats, this time AK-toting men in knee-length dresses and baggy pants waved us through the checkpoint.

One of the main roundabouts on the airport road turns directly off to the Taipan Gate, where I had taken the Japanese diplomats. It was also the most accessible entrance for us. But I watched in dismay as the convoy sped right past the Taipan turn-off.

Dammit! Was this another bluff? Were they driving us out of town to shoot us?

Then it dawned on me. They were taking us to the Abbey Gate – the same one they had repeatedly warned us was too dangerous. In fact, every news bulletin was reporting that the Abbey Gate was inaccessible due to berserk crowds trying to smash their way into the airport. We had actually watched the anarchy from the rooftop of building five the day before. One of my contacts had also told me that the assortment of hastily abandoned vehicles blocking the road meant we would have to walk for at least half a mile just to reach the gate, all the while battling our way through hordes of panicked people.

The convoy started slowing down. I could see brake lights flashing, as men dressed in black and carrying AK-47s stopped us. It looked like another checkpoint, which would make sense, as the airport was barely a mile away.

Our Taliban escort in the lead vehicle got out and walked towards the checkpoint guys, who were shoving civilians out of the way. I could hear raised voices – not quite shouting, but certainly lively enough to know the men in black were not happy. Their leader repeatedly gestured with his weapon towards us as he spoke. Whoever these guys were, they did not like us one bit – we could tell that by their hostile body language. They also seemed a lot different from the Taliban who were escorting us, their all-black clothing and aggressive attitude suggesting that they were in charge of the static checkpoints. If so, were their orders different from those of our escort? What was to stop them from taking us hostage or shooting us? As far as they were concerned, we should have been long gone out of the country.

My team and I anxiously watched the stand-off get more and more animated. Wildly gesticulating hands and loud voices are never good omens. I forcibly controlled my pounding heartbeat, mentally bracing myself and expecting to be ordered off the bus and lined up to be shot at any minute. If the men in black opened fire, we would have to run for the gate. I doubted any of us would get far.

Then the Taliban driver in charge of our escort stepped back into the front Toyota Hilux. He spun the vehicle around, facing the way we had come. It appeared the two commanders at loggerheads had come to some sort of an agreement. All of us exhaled a collective sigh of utmost relief as the convoy drivers executed U-turns and headed back, this time taking the correct turn-off towards the Taipan Gate.

The convoy stopped outside the gate. Across the street and blocking us was a small group of people, maybe fifteen to twenty, but I doubted that would be an issue with our escort. The Taliban didn't exactly hold back when dealing with civilians.

Once again, we waited like sitting ducks. After twenty minutes or so, seeing these coaches packed with foreigners, the size of the crowd started to grow, alarmingly. So much so that the Taliban had to get out of the cars and start forcibly shoving people back. Eventually, they fired shots into the air, which had the desired effect, temporarily scattering the hordes.

Why weren't we simply let in? Why weren't the gates opened for us?

The mystery deepened further when, without any explanation, our Taliban escort suddenly jumped out of the coaches, got into their trucks and hastily sped off.

Project manager Rob was as confused as I was. 'What the hell is going on?'

'No clue, mate,' I replied.

Trev stood. 'I'm bursting. Off for a piss, lads.'

He opened the door, and a welcome breeze cooled the steaming sauna of the bus interior.

'Find out what the hold-up is, will you?' I shouted after him.

Much as I hated being under the control of the Taliban, without them we would soon have serious problems with the rapidly increasing crowd swarming around the buses. The seething resentment of terrified people watching Westerners being escorted to the relative safety of the airport while they were facing unknown horrors was reaching fever-pitch. Understandably so. We had been so focused on our own survival that we hadn't thought much of what America's haphazard withdrawal actually meant for ordinary Afghans. We were now witnessing it first hand. It was distressing beyond belief.

With his bladder empty, Trev got back into the bus and said he had found out why our escort had disappeared so abruptly.

'It seems Tali are pissed off that HQ was left open and unattended, and so were told to go back.'

No one was quite sure what that meant. But it was true the compound was wide open and there was a huge amount of booty inside.

'Well, if they look under my mattress, the sling one of them wanted is on my M4,' I joked.

'You might want to ring them and ask for it back, H, if this crowd gets bigger,' said Bry.

He was right. The semi-hysterical crowd was rapidly closing in on us. As almost all other Western civilians had left several days or more ago, our convoy was seen, literally, as the last-chance saloon of escape. Many were shouting at us in broken English. The problem was that if we replied, or even nodded or smiled, they would take that as an invitation to jump aboard the buses and we would be overrun. There was nothing whatsoever we could do. I stared straight ahead, looking at the Taipan Gate and willing it to open.

A woman crossed over the road to my right. She clutched papers in one hand and cradled a small child in the other. We locked eyes for a split second and I instantly regretted it. For her, that fleeting connection was enough. She rushed up to my side window.

'Please, mister!' She held up the child. At a guess, I would say it was six months old, wrapped in a blue cotton sheet. The papers must have been the birth certificate.

'Please, mister. Please, mister.'

She began tapping the glass. I wished I could have been somewhere else. Anywhere. Just away from her pleading eyes boring into me like lasers, as she begged for help. Feeling absolutely wretched, I looked away.

She then tried to pass the tiny child through a small top window that was partially open above my head. One of the lads behind me stood and slid it shut.

Even with the window closed, I could hear her despairing voice. 'Please, mister. Please, mister.'

Those words will haunt me for the rest of my life.

The delay continued, dragging on for two long hours with more than 400 people sweltering in minibuses surrounded by a severely traumatised crowd. Obviously, the instruction not to open the gate came from inside the airport, but our presence outside was a potential spark to an extremely volatile tinderbox. Unless our buses were let in soon, the Taipan Gate was in danger of being breached. If that happened, we would have repeat scenarios of stampeding hordes trying to hang onto wheels of departing planes, or worst-case scenarios of soldiers being forced to fire into mobs of desperate civilians.

'Come on, man.' I felt like shouting at the faceless bureaucrats delaying us.

Then word came down from the lead vehicle. Everyone must be on their assigned bus and we were moving to a search area. We got ready, but still the gates remained closed. The cliché 'so near yet so far' could not have been more apt, and morale in the convoy hit rock bottom. Would we ever get out of this hellhole? I had my doubts.

Eventually, I heard the sweet sound of steel on steel. Was that the gates? I hardly dared to believe it, but finally it was happening. The heavy metal barriers slid open and the drivers engaged gears. I took one last look out the window. It was a perfect storm of absolute anguish, people wailing and pleading as we passed them. Then as the gates closed and we drove off, the clamour of the crowd faded. There were no more heartbreaking cries of 'please, mister'. No women holding up their babies.

It was quiet. We had abandoned them.

I should have been elated that at last I was getting out. I should have felt a sense of total relief that we had survived, that the nightmare of the last seven days was over.

Instead, I felt a sense of shame. I felt like a complete and utter failure. Thousands of people had died or been grotesquely maimed in this twenty-year war. Thousands of innocent people had been blown up by suicide bombers and rampaging gunmen. An entire nation had seen their country ripped apart.

Now we were tucking our tails between our legs and making a run for it. In my heart, I knew this was wrong. And that I was a part of it.

A few moments later our convoy stopped at an open car park. Several guys in plain clothes, with body armour and assault rifles, were waiting as we got out of the bus.

'All right, lads,' said one, greeting us. I don't think I had ever been so grateful to hear a British accent. They were special forces guys, who explained that they would process us before taking us to the RAF boarding teams who would put us onto a flight.

I looked over to where the Gurkhas were lining up to be cleared, and got a smile out of one or two of them. At least we had done them right. Also, to my relief, I saw the French doctor, her students and the literature teacher in the queue. They, too, were still with us. They would have a story to tell their grandkids, all right.

I snuck off to ring Nicky and let her know it was over. Well, almost, but at least we were in the airport, not manning rooftops, outnumbered and outgunned. The relief was overwhelming, and although we didn't chat for long, Nicky was not far from tears. That's saying something, as she is as brave as they come.

I strolled back into the hangar and Rob called me over.

'They're taking us to the airfield now, H. There is a plane in two hours to Dubai and then we're off to Brize Norton.'

A line of vehicles pulled up outside the special forces hangar, enough to move all of us in one go to the RAF boarding centre. I was in the lead car, and the driver pointed to the rooftops, where I could see snipers and spotters working in pairs.

'Americans,' he said. 'They're up there, day and night now.'

I nodded. Just like we had been on the roofs of buildings five and six for the past week.

As we drove to the boarding centre, I looked around, staggered at the shambolic mess that just days previously had been a bustling international airport. It was unrecognisable from the place where I had so recently dropped off the Japanese diplomats. It was little more than a massive rubbish tip, with evacuees and refugees dumping everything they owned where they stood.

Abandoned military equipment was also piled in scrapheaps – so much so that our driver had to weave around them. Armoured cars, body armour, operational kit and ballistic helmets were strewn on the ground for the victorious Taliban fighters to claim.

Pointing to all the abandoned gear, I said, 'It's going to be like Christmas once you guys pull out.'

Indeed. In a few days' time the Taliban would be the most lavishly equipped ex-guerrilla army in the world.

The drivers dropped us off at a brown tent where a handful of Afghan families sat on a kerb, cradling small children on their knees. I assumed they would be flying out, too, seeing they had made it inside the airfield. An RAF staffer came out to meet us, dressed in green fatigues. Pointing to a scale, she said the twenty-two-pound luggage weight limit would be strictly adhered to.

I knew my bag weighed more than that, as I had already broken the zip when trying to cram more stuff into it. I gingerly placed it on the scales.

'Ooohhh,' sighed my team, laughing as the scale reading flashed up. Well over! Of all the things I had to bin, the one I regretted most was a Leatherman multi-tool for a team member on leave who had specifically asked me to try to get it out. He'd had it for his entire military career, and it had huge sentimental value. Sorry, mate.

Then a call came through. 'You guys on the 14:20 come through here and get logged in on the system.'

We were going, at last. We shuffled along a corridor and into a room with two desks. Behind one sat the passenger controller and at the other was an Afghan child, drawing with some pens and paper she had been given. She looked about eight years old. She was alone.

'Passport,' asked the RAF controller. As I handed it over, I asked, 'Who's the girl?'

'She was wandering around the airfield and we can't find her parents. So she may well have been thrown over the wall.'

I could not imagine being so desperate that I would throw Lexi, my own little girl, over an airport wall into the unknown and probably never see her again. It was unthinkable.

'Jesus,' I said, shocked. 'Is she staying put or is she flying?'

The controller shrugged. Such a decision was above his pay grade. But that sad image of an abandoned girl drawing pictures was in essence a snapshot of Afghanistan's reality. It shook me to the core, magnifying the sense of failure within me. The elation of getting out of this hellhole was once again spoiled by a gnawing sense of shame.

As we waited, I started to doze off. My body was gradually accepting that I was safe. As a soldier, I could sleep anywhere, and a hard bench in an airport that was now little more than a refugee centre was as good as anywhere else.

Someone nudged me in the ribs. 'That's us. Coach is taking us to the plane.'

The Airbus A400M Atlas was on the runway. I sat up, unsure as to how much time had passed since dozing off, and glanced down at my G-Shock. It was dead. The watch that had been my faithful companion on my wrist since 2012 was a grey, blank screen. It was incredible that it had chosen to die at the exact moment I was about to board a plane to freedom. Symbolic or not, it was still a sobering reminder that my time had very nearly been up. Or at least it had certainly felt that way during the last few days and nights.

We walked out of the air-conditioned shelter and the blistering heat hit us like a blowtorch. Most of us were wearing denim jeans to look like civilians, and getting into the transport bus was like stepping into a hothouse.

'Jeans were a big mistake,' said Jay the medic, sweat starting to stream down his face.

'Yeah,' said Nath. 'But I'm even more surprised that Tali didn't execute you for wearing that shit top.'

We headed towards the plane, laughing and piss-taking as always. We were a team again.

I could see the props of the freedom bird spinning as the driver opened the bus doors. Just ahead of me was an Afghan mother and three small children. They were looking around in bewilderment. It must have been pretty frightening for them, all that noise and activity, surrounded by heavily armed men and getting onto a military plane. I smiled at the youngest as we boarded, but she stared back blankly. Probably because I looked like hell.

The plane taxied towards the furthest end of the runway, then slowly turned. As the pilot gunned the engines, I peered out of the window, watching the buildings that dotted the mountainside shoot past faster and faster, until they became a blur.

The front end lifted. Moments later we soared into the clouds.

It was a day I had thought I would not live to see.

* * * *

The pilot announced our descent into Brize Norton, the massive RAF airbase in Oxfordshire. The extraordinarily chaotic evacuation of Kabul was huge news and journalists were scrambling around to interview the last groups of non-military stragglers leaving the city. In other words, people like us and the Afghan families who had gotten out in those final hours. So we were expecting a possible media bunfight.

This posed a problem for me and my guys. Security contractors work in the shadows. Our lives, and our clients' lives, depend on it, so anonymity is crucial. As a team, we had already agreed at the stop-over in Dubai that we would not be talking to any media once we landed, either collectively or individually. Once through customs and on UK soil, we sneaked out of the building as unobtrusively as we could.

Nicky knew I would be landing that morning. It was bad timing for her, as her family were going to a wedding that day. She opted to stay home and wait for me, then we would attend the evening function if I felt up to it. As knackered as I was, I knew how much her family meant to her so agreed to go. Besides, I needed a drink!

A friend picked me up at Brize Norton. The base is a little more than two hours' drive from Lincoln and as three of the team lived near one another, we shared a lift. The journey was subdued. We had loads to say to one another about our adventures of the past week and more, but now was not the time.

I was the first stop-off and asked to be dropped at the top of the road, choosing to walk into my estate with the pitifully few possessions I had been allowed to take out of Afghanistan. I felt nervous approaching our house. Although riddled with worry, Nicky had kept the events of the last harrowing week hidden from the children and I didn't know what to expect.

As I opened the door, Nicky was there. From under her feet came Jax, shouting, 'Daddy!' The little man wrapped his arms around my leg as Lexi grabbed me around my waist. Nicky placed her head on my shoulder and I pulled her close. I could tell she was emotional but she would never let the kids see how upset she had been. She only allowed them to see her strong and happy – that's how fantastic a mother she is.

Then Jax, bursting with excitement, said, 'Daddy, let's race on Mario Kart!'

Nicky looked at me, rolling her eyes as if to say we had bigger things to deal with. But keeping things normal had to come first. I raced Jax several times, but my eyes were so heavy that concentrating on the screen took a supreme effort.

Lexi began telling me about the latest book she was reading. An avid reader, she would race through eight or nine books whenever I was away. She had just finished the latest David Walliams novel and was filling me in on all the funny bits as I dodged obstacles and raced Jax.

'Right, we're now making a cup of tea,' Nicky announced, as I lost to Jax yet again. Lexi had her head back in her book, so both the children were distracted as the novelty of my return started to wear off.

I followed Nicky into the kitchen. Her back was to me, head down, having switched on the kettle. It never took much to read her body language. For the past week, despite non-stop news broadcasts about awful events in Kabul, she had remained steadfast, keeping her house in order and our children protected. She is a true warrior. For a moment I didn't move, staring at her in awe.

She turned around, and as I held her, her body started to shake. Tears streamed down her face. We didn't say a word, just standing, holding each other. I knew this had been as big a battle for her as it had been for me. Since meeting her more than seventeen years ago, she had never questioned my work, never moaned about me being away and never once asked me to quit, even when I would come back shaken from friends' funerals.

I held her tighter still.

'It's over,' I whispered.

She looked at me; her big green eyes were red and tired. But I could see the happiness within.

Acknowledgements

Brigette Hughes (Grandma), for always being there for me; Maggie and Bob, for always being nice to me and treating me like a son, and for being wonderful grandparents; Simone and Chris, for your support and guidance when things were at their worst; Grace, for the vote of confidence when I needed it most; Shirley and Stewart Lilley, for raising me in a home from home, thank you – you are both wonderful people. Adam, for a lifetime of friendship, still going strong; Shaun, for making me commit to this book; Atila for keeping me in the fight with horrendous CrossFit sessions; Will Webb, for the friendship and shared love of motorbikes; Tony, for putting up with me as a neighbour and having to live with me ever since! All of my teammates throughout the years, it has been a pleasure to know you, work alongside you, lead you and learn from you. Thank you, all.

Special Mentions

Shaun Barber, it has been a rough ride recently, but, as always, that fighting spirit of yours shines through. You showed me back then, as you do now, that you are never out of the fight, a gentleman and a warrior to the bitter end. I am proud to have saddled up with a man of your calibre in charge.

Paul Moynan, you are a beacon of hope. Your guidance and help have been fantastic. I am proud to call you a brother and to have had you as a team member in Ramadi, watching out for me now as you did back then, a million thank-yous. We have a genuine bond, forged while having the best of times in the worst of places.

Mark McCormack, I know Luke's death hit you hardest of all the team. Hopefully his legacy lives on here. Thank you for being there for me when I had a wobble; it was always your voice and enthusiasm that brought me back to writing this. You are an inspiration to us all. Having you at the embassy made life fun, and working alongside you was an honour.

Matt Arundel, for getting me through the thirty-miler, and a lifetime of friendship ever since. We began Commando training all those years ago. Now, as then, you are a diamond.

Mick Dawson, for putting me in touch with your people and hearing out the story written here, thank you so much. If you have not read Mick's book, *Never Leave a Man Behind*, I can very highly recommend it. (Mick is also the author of *Rowing the Pacific*.)

Duncan Proudfoot and all at Ad Lib, thank you for all your hard work, for pushing and fighting for this book to be published and providing me with all the support in the world. It has been a pleasure working with you all.

Graham Spence, for sewing it all together so beautifully, a million thank-yous. From the minute we met, I knew in my heart you were the guy to bring this story to light. Your passion and eye for detail are incredible. Thank you so much for working with me on this.